# *Allies and Enemies*

# *Allies and Enemies*

## *How the World Depends on Bacteria*

Anne Maczulak

Vice President, Publisher: Tim Moore
Associate Publisher and Director of Marketing: Amy Neidlinger
Acquisitions Editor: Kirk Jensen
Editorial Assistant: Pamela Boland
Operations Manager: Gina Kanouse
Senior Marketing Manager: Julie Phifer
Publicity Manager: Laura Czaja
Assistant Marketing Manager: Megan Colvin
Cover Designer: Alan Clements
Managing Editor: Kristy Hart
Senior Project Editor: Lori Lyons
Copy Editor: Geneil Breeze
Proofreader: Apostrophe Editing Services
Senior Indexer: Cheryl Lenser
Compositor: Nonie Ratcliff
Senior Manufacturing Buyer: Dan Uhrig


Publishing as FT Press
Upper Saddle River, New Jersey 07458

FT Press offers excellent discounts on this book when ordered in quantity for bulk purchases or special sales. For more information, please contact U.S. Corporate and Government Sales, 1-800-382-3419, corpsales@pearsontechgroup.com. For sales outside the U.S., please contact International Sales at international@pearson.com.



Printed in the United States of America

First Printing July 2010

ISBN-10: 0-13-701546-1
ISBN-13: 978-0-13-701546-7

Pearson Education LTD.
Pearson Education Australia PTY, Limited.
Pearson Education Singapore, Pte. Ltd.
Pearson Education North Asia, Ltd.
Pearson Education Canada, Ltd.
Pearson Educación de Mexico, S.A. de C.V.
Pearson Education—Japan
Pearson Education Malaysia, Pte. Ltd.

*Library of Congress Cataloging-in-Publication Data*

Maczulak, Anne E. (Anne Elizabeth), 1954-
Allies and enemies : how the world depends on bacteria / Anne E. Maczulak.
p. ; cm.
Includes bibliographical references and index.
ISBN-13: 978-0-13-701546-7 (hardback : alk. paper)
ISBN-10: 0-13-701546-1 (hardback : alk. paper) 1. Bacteria—Popular works. 2. Microbial biotechnology—Popular works. 3. Microbiology—Popular works. I. Title.
[DNLM: 1. Bacteria. 2. Bacterial Physiological Phenomena. 3. Bacteriology—history. QW 50 M177a 2010]
QR56.M26 2010
579.3—dc22

2010006589

# Contents

# Acknowledgments

I became a microbiologist in Burk A. Dehority's laboratory in 1978 studying anaerobes in cattle, sheep, and horses. From that point on I have met or worked with some of the most respected researchers in the fields of anaerobic, environmental, and water microbiology. I'm sure they have forgotten more microbiology than I ever learned, but we collectively must admit that bacteria still hold a vast world of unknowns. I thank all of my professors of microbiology at the Ohio State University and the University of Kentucky.

For this book I owe thanks to Bonnie DeClark, Dana Johnson, Priscilla Royal, Sheldon Siegel, Meg Stiefvater, and Janet Wallace for their advice on chapter content. Special gratitude is due Dennis Kunkel and Richard Danielson who always seem to offer encouragement when it is needed the most. Thanks are due to Amanda Moran and Kirk Jensen for their valuable guidance, and to Jodie Rhodes for tireless encouragement and support.

# About the Author

**Anne Maczulak** grew up in Watchung, New Jersey, with a plan to become either a writer or a biologist. She completed undergraduate and master's studies in animal nutrition at The Ohio State University, her doctorate nutrition and microbiology from the University of Kentucky, and conducted postdoctoral studies at the New York State Department of Health. She also holds an MBA from Golden Gate University in San Francisco.

Anne began her training as a microbiologist studying the bacteria and protozoa of human and animal digestive tracts. She is one of a relatively small group of microbiologists who were trained in the Hungate method of culturing anaerobic microbes, meaning microbes that cannot live if exposed to oxygen. In industry, Anne worked in microbiology laboratories at Fortune 500 companies, developing anti-dandruff shampoos, deodorants, water purifiers, drain openers, septic tank cleaners, and disinfectants—all products that relate to the world of microbes. She conducted research in the University of California-San Francisco's dermatology group, testing wound-healing medications, antimicrobial soaps, and foot fungus treatments.

In graduate school, other students and a few professors had seemed nonplussed when Anne filled her elective schedule with literature courses. Anne was equally surprised to learn that so many of her peers in science found pursuit of the arts to be folly. In 1992, with more than a decade of "growing bugs" on her resume, she packed up and drove from the east coast to California to begin a new career as a writer while keeping microbiology her day job. And yes, it was possible to be both a writer and a scientist.

While toiling evenings on a mystery novel set in a microbiology lab, Anne continued working on various laboratory projects intended either to utilize good microbes or eliminate deadly ones. A decade later, Anne began her career as an independent consultant and has successfully blended writing with biology. Although the mystery novel never made it off the ground, Anne has since published ten books on microbes and environmental science. She focuses on making highly technical subjects easy to understand. From her unique perspective, Anne inspires her audiences into wanting to know more about microbes, and perhaps even like them.

# Introduction

In the mid-1600s, Europe's population had been decimated by three centuries of bubonic plagues. The deadliest had been the Black Death, killing one-third of the population between 1347 and 1352. Between each epidemic European cities repopulated and rebuilt their commerce. In Amsterdam, the Dutch had ceded dominance of the seas to England but retained a central role in European finance and the trade routes. Glass, textiles, and spices moved by the ton through the Netherlands' ports.

After apprenticing in Amsterdam, cloth merchant Antoni van Leeuwenhoek returned to his birthplace Delft to start his own business and capitalize on the growing economy. Needing a way to assess fabric quality and compete with established clothiers, van Leeuwenhoek experimented with glass lenses of various thicknesses to magnify individual threads. More than 75 years earlier, eyeglass makers Zacharias Janssen and his father, Hans, had put multiple lenses in sequence to amplify magnification and in doing so invented the first compound microscope. Van Leeuwenhoek used mainly single lenses, but he formed them with precision, enabling him to observe the microscopic world as no one had before.

Van Leeuwenhoek continued tinkering with new microscope assemblies and word spread of the clever new invention. More for hobby than for science, he studied various items from nature. Using a magnification of 200 times, van Leeuwenhoek spied tiny objects moving about in rainwater, melted snow, and the plaque sampled from teeth. He described the microscopic spheres and rods in such detail that scientists reading his notes three centuries later would recognize them. Van Leeuwenhoek called the minute creatures "animalcules" and introduced the first studies of the microscopic world. The animalcules would someday be known as bacteria, and van Leeuwenhoek would be credited with creating the science called microbiology.

Bacteria are self-sufficient packets of life, the smallest independently living creatures on Earth. Although bacteria derive clear benefits from living in communities, they do well in a free-living form called the planktonic cell. Bacteria as a group are not bound by the constraints that marry protozoa to aqueous places, algae to sunshine, and fungi to the soil.

The key to understanding microbes is to understand the cell. A cell is the simplest collection of molecules that can live. Life can be harder to define. Life has a beginning, an aging process, and an end, and during this span it involves reproduction, metabolism, and some sort of response to the environment. Biologists think of cells as the most basic unit of life in the way that an atom is the basic unit of chemistry.

Microbiology encompasses all biological things too small to be seen with the unaided eye. Mold spores, protozoa, and algae join bacteria in this world, each with attributes that would appear to give them advantages over the other microbes. Mold spores, for instance, are hardy, little spiked balls that withstand drought and frost and travel for miles on a breeze. Many bacteria do something similar by forming a thick-walled endospore that can outlast a mold spore by centuries. Protozoa meanwhile stalk their nutrition, which often comes in the form of bacteria. Why hunt a hundred different nutrients when you can swallow one bacterial cell for dinner? But bacteria roll out their own version of predation. Certain bacteria form cooperative packs that conserve energy as they roam their environment, searching for other bacteria to eat. Finally, algae appear to hold an ace because they produce their own food by absorbing solar energy and using it to power photosynthesis. But bacteria rise to the challenge here, too. Some bacteria live cheek-by-jowl with algae at the water's surface and carry out the same photosynthesis. Other bacteria exist at greater depths and use the scarce light rays that filter through the water's surface layer. Give bacteria the power of speech and they might say, "Anything you can do I can do better."

Bacteria as a group live everywhere, reproduce on their own without the need for a mate, and depend on no other cells for their survival. Unlike any other type of cell in biology, bacteria do these things using the simplest cell in biology. What about viruses, which are often described as the simplest biological beings in existence? The science of microbiology has adopted viruses mainly because viruses are microscopic and biological. But viruses cannot perform all the functions that would make them a living thing: a life cycle, metabolism, and interaction with the environment. Viruses depend entirely on living cells for their survival. A single virus particle dropped into even the most comfortable environment would be a lifeless speck with no capabilities of its own.

Various theories have been put forth to explain the origin of viruses in relation to bacteria. Viruses may have descended from a primitive form of nucleic acid, meaning deoxyribonucleic acid (DNA) or ribonucleic acid (RNA). RNA carries information inside cells just as DNA carries genes. RNA interprets the code in DNA's genes and uses this information to assemble cellular components. RNA would be a likely candidate for originating viruses because its structure is simpler than DNA's; DNA contains two long chains that make up its molecule and RNA has only one chain. Perhaps ancient RNA directed the early processes of building more complex molecules such as a nucleic acid wrapped in protein, the basic structure of a virus. (A protein is a long strand of amino acids folded into a specific shape.) A second contrasting theory views viruses as self-replicating pieces of RNA or DNA cast out from early bacteria. The pieces somehow became enveloped in protein and thus turned into the first virus. Microbiologists have also considered a scenario in which evolution reversed and bacterial cells regressed by shedding much of their cellular structure until only nucleic acid surrounded by protein remained. The theories fall into and out of favor, but one thing is certain: bacteria and viruses share a very long history on Earth.

Fungi, protozoa, algae, plants, and all animal life, including humans, belong to the Domain Eukarya. The cells that make up eukaryotes have internal structures called organelles. The organelles give eukaryotic cells an orderliness that bacteria lack and help refine the basic activities of the cell: building compounds, breaking down compounds, and communicating with other cells. But managing a lot of infrastructure also requires extra work. During cell reproduction, each organelle must be allocated to the two new cells. In sexual reproduction, a eukaryote needs another eukaryotic cell to propagate the species. Members of Domain Bacteria and bacterialike microbes in Domain Archaea split in half by binary fission without the worries of managing organelles, which bacteria and archaea lack. (Archaea are indistinguishable from bacteria in a microscope, and many scientists, even microbiologists, lump the two types of microbes together.)

Before people knew of the existence of bacteria, they put bacteria to work making or preserving foods and decomposing waste. Although humanity's relationship with bacteria extends to humans' earliest history, studies of these cells began in earnest only 200 years ago, and the major discoveries in bacterial evolution emerged in the past 50 years. Bacterial genetics bloomed in 1953 when James Watson, Francis Crick, and Rosalind Franklin studied a thick, mucuslike substance from *Escherichia coli* and thus determined the structure of DNA.

Bacteriology required microscopes to improve before this science could advance. Van Leeuwenhoek provided a starting point, but others refined the instrument, particularly van Leeuwenhoek's British contemporary Robert Hooke. Hooke invented a way to focus light on specimens to make the magnified image easier to study. By the 1800s, microbes had captured the imagination of scientists and microbiology would enter a period from 1850 to the early 20th century called the Golden Age of Microbiology. By the close of the Golden Age, microbiologists had solved a number of health and industry problems related to bacteria. Microbiology's eminent Louis Pasteur would raise the stature of microbiologists to veritable heroes.

The emergence of electron microscopy in the 1940s enabled microbiologists to see inside individual bacterial cells. This achievement plus the studies on DNA structure and replication launched a new golden period, this time involving cellular genetics. By learning how bacteria control and share genes, geneticists moved beyond simply crossing red flowering plants with white. Genetics reached the molecular level. Some electron microscopes now produce images of atoms, the smallest unit of matter. With these abilities, scientists have uncovered the fine points of cell reproduction. Genetic engineering, biotechnology, and gene therapy owe their development to the first microscopic studies on cell organization.

Microbiologists also peer outward from bacterial cells to entire ecosystems. Ecologists have discovered bacteria in places no one thought a creature could live, and the bacteria do not merely tolerate these places, they thrive. Many of the surprises have come from extremophiles that live in environments of extraordinary harshness, by human standards, where few other living things can survive. Industries have mined extremophiles for enzymes that work either at extremely hot or frigid conditions. Polymerase chain reaction (PCR), for example, relies on an enzyme from an extremophile to run reactions between a range of 154°F and 200°F. PCR replicates tiny bits of DNA into millions of copies in a few hours. By using the enzyme (called restriction endonuclease) from extremophiles, microbiologists can track disease outbreaks, monitor pollution, and catch criminals.

Bacteria recycle the Earth's elements and thereby support the nutrition of all other living things. Bacteria feed us and clean up our wastes. They help regulate the climate and make water drinkable. Some bacteria even release compounds into the air that draw moisture droplets together to form clouds. But most people overlook the benefits of bacteria and focus instead on what I call the "yuck factor." "Are bacteria really everywhere?" "Is my body crawling with bacteria right now?" "Is *E. coli* on doorknobs?" The answers are yes, yes, and yes. To a microbiologist, this is a wonderful thing.

Bacteria thrive on every surface on Earth, and almost all bacteria possess at least one alternative energy-generating system if the preferred route hits a snag. And if some bacteria do not thrive, they at least develop mechanisms that allow them to ride out catastrophe. The apparent indestructibility of bacteria may fuel the fear people have toward them. We fear infectious disease, resistant superbugs, and the high mortalities that bacteria have already caused in history. Pathogens in fact make up a small percentage of all bacteria, yet if asked to name ten bacteria in 15 seconds, almost everyone would tick off the names of pathogens.

I am here to improve the public image of bacteria. Bacteria can and do harm people, but this happens almost exclusively when people make mistakes that let dangerous bacteria gain an advantage. The benefits we receive from bacteria far outweigh the harm. By understanding the wide variety of Earth's bacteria, people can put some of their fears aside and appreciate the vital contributions of these microbes. The bacterial universe may at first glance seem invisible. But as you get to know the bacteria that influence your life each day, they become easier to see even if they truly remain invisible. Bacteria have been called "friendly enemies," but I think that sends the wrong message. Bacteria are powerful friends. We will never defeat bacteria, nor do we want to. Like most friends with lots of power, it is best to respect them, treat them well, and keep them close.

# 1

## Why the world needs bacteria

What is a bacterium? Bacteria belong to a universe of single-celled creatures too small, with rare exceptions, to be seen by the unaided eye, but exist everywhere on Earth. Being small, simple, and many confers on bacteria advantages that allow them to not only survive but also to affect every mechanism by which the planet works. Bacteria influence chemical reactions from miles above the Earth's surface to activities deep within the Earth's mantle.

Bacteria range in size from *Thiomargarita namibiensis*, which reaches 750 micrometers (µm) end to end and is visible to the naked eye, to *Francisella tularensis* measuring only 0.2 µm in diameter. Since 1988, microbiologists have explored a new area involving "nanobacteria." These microbes measure 0.05 µm in diameter or one-thousandth the volume of a typical bacterial cell. Excluding these unusual giants and dwarfs, most bacteria are between 0.5 and 1.5 µm in diameter and 1 to 2 µm long, or less than one-twentieth the size of the period at the end of this sentence. The volume of bacterial cells ranges from 0.02 to 400 $\mu m^3$. One of many advantages in being small involves the ability to sense environmental changes with an immediacy that large multicellular organisms lack.

Bacterial simplicity can deceive. The uncomplicated structure actually carries out every important biochemical reaction in Earth's ecosystems. Bacteria have an outer cell wall that gives them their distinctive shapes (see Figure 1.1) and overlays a membrane, which holds in the watery cytoplasm interior and selectively takes in nutrients, restricts the entry of harmful substances, and excretes wastes. This membrane resembles the membranes of all other living things. That is, it is consists of a bi-layer of proteins and fats that communicates with the aqueous environment and confines the cell contents to the cell interior. Inside the membrane bi-layer proteins and fats line up in a way that hydrophilic or water-attracting portions of the compounds face out or into the cytoplasm, and hydrophobic compounds point into the membrane. The character of membrane fats enables them to assemble spontaneously if put into a beaker of water. The ease with which membranes assemble likely helped the first cells to develop on Earth.

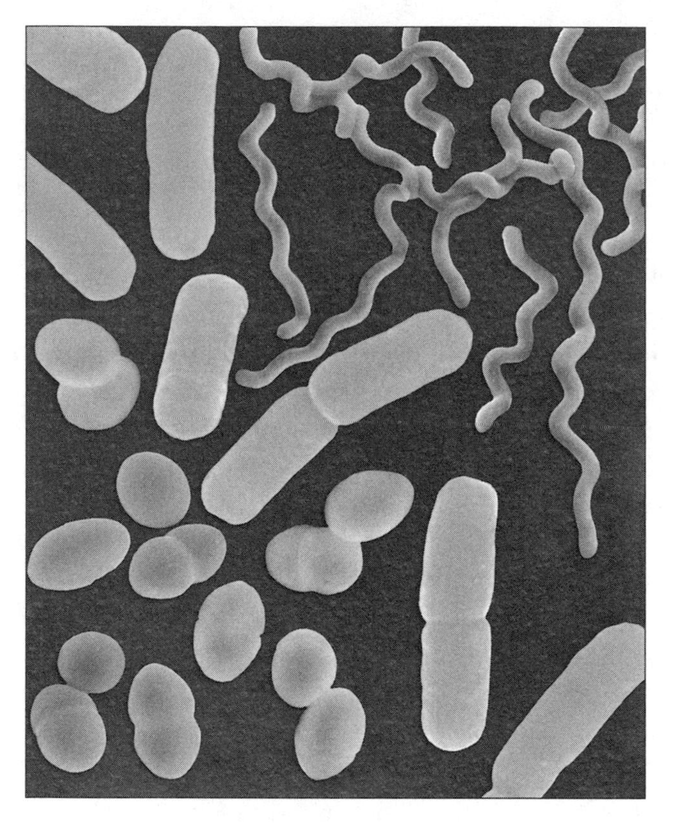

Figure 1.1 Bacteria shapes. Cell shape is hardwired into bacteria genetics. No animal life adheres as strictly to a standard shape as bacteria and algae called diatoms. (Courtesy of Dennis Kunkel Microscopy, Inc.)

The bacterial cytoplasm and membrane hold various enzymes that keep the cell alive. Bacterial deoxyribonucleic acid (DNA), the depository of information formed over the millennia, appears in the cytoplasm as a disorganized mass (seen only with an electron microscope), but it actually contains precise folds and loops that decrease the chances of damage and facilitate repair. Tiny protein-manufacturing particles called ribosomes dot much of the remainder of the cytoplasm.

Bacteria require few other structures. Motile bacteria have whiplike tails called flagella for swimming, photosynthetic cyanobacteria contain light-absorbing pigments, and magnetotactic species, such as *Aquaspirillum magnetotacticum*, contain a chain of iron magnetite particles that enable the cells to orient toward Earth's poles. These micro-compasses help *Aquaspirillum* migrate downward in aqueous habitats toward nutrient-rich sediments.

Though tiny, bacteria occupy the Earth in enormous numbers. Microbiologists estimate total numbers by sampling soil, air, and water and determining the bacterial numbers in each sample, and then extrapolating to the size of the planet with the aid of algorithms. Guesswork plays a part in these estimates. Bacteria exist 40 miles above the Earth and 7 miles deep in the ocean, and most of these places have so far been inaccessible. The total numbers of bacteria reach $10^{30}$. Scientists struggle to find a meaningful comparison; the stars visible from Earth have been estimated at "only" $7 \times 10^{22}$. The mass of these cells approaches $2 \times 10^{15}$ pounds, or more than 2,000 times the mass of all 6.5 billion people on Earth. Of these, the overwhelming majority lives in the soil.

Bacteria can stretch the limits of our imagination with small size and massive numbers. Both of these attributes help bacteria, and by the biological processes they carry out, bacteria also ensure that humans survive.

## Tricks in bacterial survival

Bacteria and bacterialike archaea survive challenging conditions through the benefit of adaptations accrued in evolution. Survival techniques might be physical or biochemical. For example, motility in bacteria serves as an excellent way to escape danger. In addition to flagella that help bacteria swim through aqueous environments, some bacteria can glide over surfaces, and others start twitching frantically to propel themselves. Certain bacterial species develop impregnable shells called endospores. Others use biochemical aids to survival to counter the effects of acids, bases, salt, high or low temperature, and pressure.

A large number of bacteria use a modified version of a capsule for protection. The cells build long, stringy lipopolysaccharides, which are polysaccharides (sugar chains) with a fatty compound attached and which extend into the cell's surroundings. The bacteria that make these appendages, called O antigens, construct them out of sugars rarely found in nature. As a consequence, protozoa that prey on bacteria do not recognize the potential meal and swim past in search of "real" bacteria.

Archaea seem to be Earth's ultimate survivors because of the extreme environments they inhabit. Archaea and bacteria both belong to the prokaryotes, one of two major types of cells in biology, the other being more complex eukaryotic cells of algae, protozoa, plants, and animals. Because archaea inhabit extreme environments that would kill most terrestrial animal and plant life, the archaea are sometimes thought of as synonymous with "extremophile." The outer membrane of archaea living in boiling hot springs contain lipid (fat-like) molecules of 30 carbons or more, larger than most natural fatty compounds. These lipids and the ether bonds that connect them stabilize the membrane at extremely high temperatures. News stories often tell of new bacteria found at intense pressures 12,000 feet deep on the ocean floor at vents called black smokers. These hydrothermal vents spew gases at 480°F, release acids, and reside at extreme pressures, so any organisms living there would truly be a news item. The organisms living near black smokers are usually archaea, not bacteria. Archaea also dominate habitats of high salt concentration, such as salt lakes, or places completely devoid of oxygen, such as subsurface sediments. Because of the difficulty of getting at many archaea and their aversion to growing in laboratory conditions, studies on archaea trail those completed on bacteria.

Some bacteria also survive in the same extreme conditions favored by archaea. The aptly named *Polaromonas* inhabits Antarctic Sea ice where temperatures range from 10°F to –40°F by slowing its metabolism until it reproduces only once every seven days. By comparison, *E. coli* grown in a laboratory divides every 20 minutes. *Polaromonas* is a psychrophile or cold-loving microbe. *Thermus aquaticus* is the opposite, a thermophile that thrives in hot springs reaching 170°F by synthesizing heat-stabile enzymes to run its metabolism. Enzymes of mesophiles, which live in a comfortable temperature range of 40°F to 130°F, unfold when heated and thus lose all activity. Mesophiles include the bacteria that live on or in animals, plants, most soils, shallow waters, and foods. The bacteria that live in harsh conditions that mesophiles cannot endure are the Earth's extremophiles.

The genus *Halococcus*, a halophile, possesses a membrane-bound pump that constantly expels salt so the cells can survive in places like the Great Salt Lake or in salt mines. Barophilic bacteria that hold up under intense hydrostatic pressures from the water above are inexorably corroding the *RMS Titanic* 12,467 feet beneath the Atlantic. These barophiles contain unsaturated fats inside their membranes that make the membrane interior more fluid than the fats in other bacterial membranes. Unsaturated fats contain double bonds between some of the carbon atoms in the chainlike fat rather than single bonds that predominate saturated fats. At pressures of the deep ocean, normal membrane liquids change into the consistency of refrigerated butter, but the special membrane composition of barophiles prevents such an outcome that would render the membrane useless. A later chapter discusses why red-meat animals store mainly saturated fats and pork and chicken store more unsaturated fats.

The acidophile *Helicobacter pylori* that lives in the stomach withstands conditions equivalent to battery acid of pH 1 or lower by secreting compounds that neutralize the acid in their immediate surroundings. Even though an acidophile lives in strong acids that would burn human skin, it remains protected inside a microscopic cocoon of about pH 7. Additional extremophiles include alkaliphiles that live in highly basic habitats such as ammonia and soda lakes; xerophiles occupy habitats without water; and radiation-resistant bacteria survive gamma-rays at doses that would kill a human within minutes. *Deinococcus*, for instance, uses an efficient repair system that fixes the damage caused to the DNA molecule by radiation at doses that would kill a human. This system must be quick enough to complete the repair before *Deinococcus's* next cell division.

All bacteria owe their ruggedness to the rigid cell wall and its main component, peptidoglycan. This large polymer made of repeating sugars and peptides (chains of amino acids shorter than proteins and lacking the functions of proteins) occurs nowhere else in nature.

Peptidoglycan forms a lattice that gives species their characteristic shape and protects against physical damage. A suspension of bacteria can be put in a blender, whipped, and come out unharmed.

Archaea construct a cell wall out of polymers other than peptidoglycan, but their cell wall plays the same protective role. Furthermore, because archaea have a different cell wall composition than bacteria, they resist all the antibiotics and enzymes that attack bacterial cell walls. This quirk would seem to make archaea especially dangerous pathogens to humans, but on the contrary, no human disease has ever been attributed to an archaean.

In a microscope, bacteria present an uninspiring collection of gray shapes: spheres, rods, ovals, bowling pins, corkscrews, and boomerangs. Microbiologists stain bacteria with dyes to make them more pronounced in a light microscope or use advanced types of microscopy such as dark field or phase contrast. Both of these latter methods create a stunning view of bacteria illuminated against a dark background.

When bacteria grow, the cell wall prevents any increase in size so that bacterial growth differs from growth in multicellular organisms. Bacteria grow by splitting into two new cells by binary fission. As cell numbers increase, certain species align like a strand of pearls or form clusters resembling grapes. Some bacteria form thin, flat sheets and swarm over moist surfaces. The swarming phenomenon suggests bacteria do not always live as free-floating, or planktonic, beings but can form communities. In fact, bacterial communities represent more than a pile of cells. Communities contain a messaging system in which identical cells or unrelated cells respond to each other and change their behavior. This adaptation is called quorum sensing.

Quorum sensing begins when cells excrete a steady stream of signal molecules resembling amino acids. The excreted signal travels about 1 μm so that neighboring cells can detect it with specific proteins on their surface. When the receptors clog with signal molecules, a cell gets the message that other cells have nudged too close; the population has grown too dense. The proteins then turn on a set of genes that induce the bacteria to change their behavior. Different types of bacterial communities alter behavior in their own way, yet throughout bacteriology communities offer bacteria a superb survival mechanism. Some communities swarm, others cling to surfaces, and yet others can cover a pond's surface and control the entire pond ecosystem.

## Bacterial communities

Swarm cells start growing like any other bacterium on laboratory-prepared nutrient medium. (Media are liquids or solids containing gel-like agar that supply bacteria with all the nutrients needed for growth.) They metabolize for a while, split in two, and repeat this until nutrients run low. Rather than halting the colony's growth, swarm cells signal each other to change the way they reproduce. The swarmer *Proteus* develops a regular colony when incubated, each cell about three µm in length. After several hours, cells on the colony's outer edge elongate to 40 to 80 µm and sprout numerous flagella. Ten to 12 flagellated cells team up and then squiggle away from the main colony. By forming teams of cells lined up in parallel, 50 times more flagella power the cells forward than if one *Proteus* headed out on its own. Several millimeters from the main colony, the swarmers stop and again begin to reproduce normally. As generations of progeny grow, they build a ring of *Proteus* around the original colony, shown in Figure 1.2. At a certain cell density in the ring, *Proteus* repeats the swarming process until a super-colony of concentric rings covers the entire surface. When two swarming *Proteus* colonies meet, they do not overrun each other. The two advancing fronts stop within a few µm of each other, repelled by their respective defenses. *Proteus* produces an antibacterial chemical called bacteriocin. The specific bacteriocin of each swarmer colony protects its turf against invasion.

Other swarmer bacteria use hairlike threads called pili rather than flagella, and cast their pili ahead to act as tethers. By repeatedly contracting, the cells drag themselves forward to up to 1.5 inches per hour. Petri dishes measure only 4 inches across, but if dishes were the size of pizzas, swarm cells would cover the distance.

Communities such as biofilm grow on surfaces bathed in moisture. Biofilms cover drinking water pipes, rocks in flowing streams, plant leaves, teeth, parts of the digestive tract, food manufacturing lines, medical devices, drain pipes, toilet bowls, and ships' hulls. Unlike swarming colonies, biofilm contains hundreds of different species, but they too interact via quorum sensing. (Bacteria that merely attach to surfaces such as skin are not true biofilms because they do not coalesce into a community that functions as a single entity.) Biofilm begins with a few cells that stick to a surface by laying down a coat of a sticky polysaccharide. Other bacteria hop aboard and build the diverse biofilm colony.

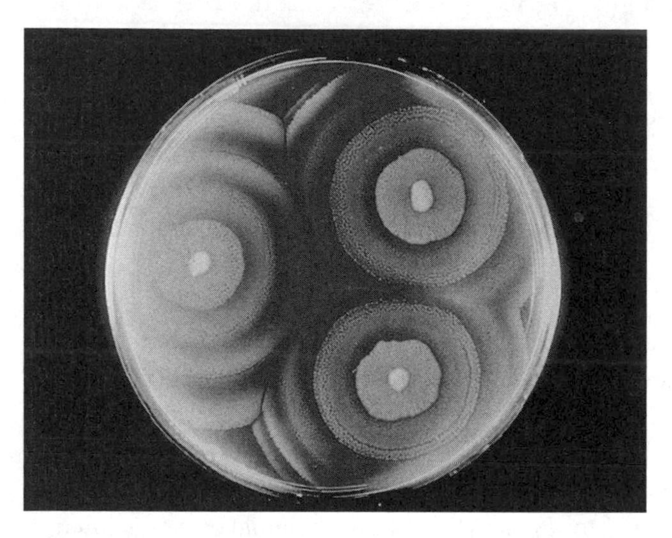

Figure 1.2 The swarming bacterium *Proteus mirabilis*. *Proteus* swarms outward from a single ancestor cell and forms concentric growth rings with each generation. (Courtesy of John Farmer, CDC Public Health Image Library)

Biofilms facilitate survival by capturing and storing nutrients and excreting more polysaccharide, which protects all the members against chemicals such as chlorine. Eventually fungi, protozoa, algae, and inanimate specks lodge in the conglomeration of pinnacles and channels. When the biofilm thickens, signals accumulate. But because many different species live in the biofilm, the signals differ. Some bacteria stop making polysaccharide so that no more cells can join the community. The decrease in binding substance causes large chunks to break from the biofilm, move downstream, and begin new biofilm. (This constant biofilm buildup and breakdown causes great fluctuations in the number of bacteria in tap water. Within a few hours tap water can go from a few dozen to a thousand bacteria per milliliter.) Meanwhile, other bacteria ensure their own survival by increasing polysaccharide secretion, perhaps to suffocate nearby microbes and reduce competition.

Pathogens likely use similar strategies in infection by turning off polysaccharide secretion. With less polysaccharide surrounding the bacteria, the cells can reproduce rapidly. Then when pathogen numbers reach a critical level in the infected area, polysaccharide secretion returns to quash competitors.

A second type of multispecies community, the microbial mat, functions in complete harmony. Microbial mats lie on top of still waters and are evident by their mosaic of greens, reds, oranges, and purples from pigmented bacteria. Two types of photosynthetic bacteria dominate microbial mats: blue-greenish cyanobacteria and purple sulfur-using bacteria. During the day, cyanobacteria multiply and fill the mat's upper regions with oxygen. As night falls and cyanobacteria slow their metabolism, other bacteria devour the oxygen. Purple bacteria prefer anoxic conditions, so they live deep in the mat until the oxygen has been depleted. In the night, the purple bacteria swim upward and feast on organic wastes from the cyanobacteria. The sunlight returns, and the purple bacteria descend to escape the photosynthesis about to replenish the upper mat with oxygen. As they digest their meal, these bacteria expel sulfide compounds that diffuse to the top layer. There, sulfur-requiring photosynthetic bacteria join the cyanobacteria (and some algae) in a new cycle. An undisturbed mat literally breathes: absorbing oxygen and emitting it, expelling carbon dioxide and inhaling it one breath every 24 hours. Microbial mats' diurnal cycle makes them a distinctive microbial community.

Communities are mixtures of species within an ecosystem. Ecosystems contain living communities that interact with the nonliving things around them: air, water, soil, and so on. Bacteria participate in every phase of ecosystem life, but to learn about bacteria microbiologists must remove them from the environment and study one species at a time in a laboratory. A collection of bacterial cells all of the same species is called a population, or in lab talk a pure culture.

Microbiologists learn early in their training the tricky job of keeping all other life out of a pure culture by using aseptic technique. Aseptic—loosely translated as "without contamination"—technique requires that a microbiologist manipulate cultures without letting in any unwanted bacteria. They accomplish this by briefly heating the mouth of test tubes over a Bunsen burner flame, similarly flaming metal inoculating loops, and learning to keep sterilized equipment from touching unsterilized surfaces. Surgeons follow the same principles after they scrub up for surgery.

## Under the microscope

For the two centuries following van Leeuwenhoek's studies, microscopes improved, but microbiologists still needed a way to distinguish cells from inanimate matter in a specimen. They tested a variety of chemical dyes on bacteria with usually unsatisfactory results. In 1884, Danish physician Hans Christian Gram formulated through trial and error a stain for making bacteria visible in the tissue of patients with respiratory infection. On a glass slide, Gram's recipe turned some of the bacteria dark purple and others pink. The new method served Gram's purposes for diagnosing disease, but he had no notion of the impact the Gram stain would have on bacteriology.

The Gram stain divides all bacteria into two groups: gram-positive and gram-negative. This easy procedure serves as the basis for all identifications of bacteria from the sick, from food and water, and from the environment. Every student in beginning microbiology commences her education by learning the Gram stain.

Bacteria with thick cell walls of peptidoglycan retain a crystal violet-iodine complex inside the wall. These cells turn purple and are termed gram-positive. Other species cannot retain the stain-iodine complex when rinsed with alcohol. These gram-negative cells remained colorless, so Gram added a final step by soaking the bacteria in a second stain, safranin, that turned all the colorless cells pink. All bacteriologists now use the Gram stain as the first step in identification, monitoring food and water for contamination, and diagnosing infectious disease.

In the more than 100 years since Gram invented the technique, microbiologists have yet to figure out all the details that make some cells gram-positive and others gram-negative. The thick peptidoglycan layer in gram-positive cell walls has an intricate mesh of cross-links. This structure acts as a net to retain the large crystal violet-iodine aggregate and might keep the alcohol from reaching the stain and washing it out. By contrast, the gram-negative cell wall is more complex. The thin peptidoglycan in gram-negatives lies in between membranes on both the outer and inner surfaces of the cell. The thinness of the layer has been proposed as one reason why gram-negative cells cannot hold onto the stain.

Few hard and fast rules can be attributed to gram-positive and gram-negative populations. Gram-negative bacteria were once thought to be more numerous than gram-positives and have a higher proportion of pathogens, but these generalizations probably hold little merit. The Gram reaction nevertheless helps gives clues to microbiologists about potential trouble. Food, water, consumer products such as shampoo, and skin with high concentrations of gram-negative bacteria signal possible fecal contamination. That is because *E. coli* and all other bacteria in its family come from animal intestines. But gram-positive bacteria are not totally benign. Gram-positive bacteria recovered from a person's upper respiratory tract might indicate strep throat (from *Streptococcus*) or tuberculosis. Skin wounds infected with gram-positives range in seriousness from *Staph* infections (from *Staphylococcus*) to anthrax. In the environment, the known gram-negative and gram-positive species distribute almost evenly in soils and waters.

During the time Gram worked out his new procedure, German physician Walther Hesse left his job of ten years tending to uranium miners in Saxony who were dying of lung cancer (although the disease had not yet been identified). After two years in Munich working in public hygiene, he became an assistant to Robert Koch who was second only to Louis Pasteur as the world's eminent authority on microbes. Originally a country doctor in a small German village, Koch had already immersed himself in the behavior of anthrax and tuberculosis bacteria in test animals. From these studies he began developing a procedure for proving that a given bacterial species caused a specific disease. In 1876, Koch established a set of criteria that a bacterium must meet in test animals to be identified as the cause of disease. The criteria to become known as Koch's postulates laid the foundation for diagnosis of infectious disease that continues today.

Medical historians have debated whether the criteria attributed to Robert Koch should be called the Henle-Koch postulates. Koch received his early training under German physician Jacob Henle who in 1840 published a list of criteria for confirming the cause of infectious disease. The criteria proposed by Koch were similar to Henle's, but the origin of Koch's postulates probably came by a gradual evolution of ideas with each new experiment on pathogens. I explain Koch's postulates here:

1. The same pathogen must be present in every case of a disease.
2. The pathogen must be isolated from the diseased host and grown in a laboratory to show it is alive.
3. The pathogen should be checked to confirm its purity and then injected into a healthy host (a laboratory animal).
4. The injected pathogen must cause the same disease in the new host.
5. The pathogen must be recovered from the new host and again grown in the laboratory.

Some bacteria do not conform to Koch's postulates. For example *Mycobacterium tuberculosis*, the cause of tuberculosis, also infects the skin and bones in addition to the lungs. *Streptococcus pyogenes* causes sore throat, scarlet fever, skin diseases, and bone infections. Pathogens that cause several different disease conditions can be difficult to fit into the criteria for diagnosing a single disease.

In developing these criteria, Koch made another contribution to the fundamentals of microbiology by introducing a way to obtain pure cultures. For Koch's postulates to work, a microbiologist needed a pure culture of the potential pathogen. Without bacteria in pure form, no one would be able to prove bacterium A caused disease A, bacterium B caused disease B, and so forth. Koch used potato slices for growing bacterial colonies and for his studies used only colonies that were isolated from all other colonies. This concept seems elementary today, but it helped microbiologists of Koch's time rid their experiments of contaminants. To this day, prominent researchers have reported results only to make an embarrassing retraction months later because all of the data were collected on a contaminant.

When Hesse joined Koch's laboratory, Koch had stopped using potato slices and substituted gelatin as a handier surface for growing pure colonies. Soon both men were grousing about gelatin's flaws. In hot summers, the gelatin turned to liquid. Most other times, protein-degrading bacteria turned it into a useless blob. Hesse's wife, Angelina, often came to the lab to help—this was a period in Germany when women were taking their first steps into professions. Lina, as Hesse called her, was an amateur artist and helped Koch and Hesse by drawing the bacterial colonies they had grown in the laboratory. She soon understood why the two microbiologists needed something better than gelatin. Lina suggested that they try agar-agar, a common ingredient at the time for solidifying puddings and jellies. Wolfgang, the Hesses' grandson recalled in 1992, "Lina had learned about this material as a youngster in New York from a Dutch neighbor who had immigrated from Java." People living in the warm East Indian climate noticed that birds gathered a substance from seaweed and used it as a binding material in nests. The material did not melt and did not appear to spoil—bacteria cannot degrade it.

Hesse passed on to Koch the idea of replacing gelatin with agar-agar. Koch immediately formulated the agar with nutrients into a medium that melted when heat-sterilized and solidified when cooled (see Figure 1.3). Koch published a short technical note on the invention but mentioned neither of the Hesses. Lina lived for 23 years after her husband's death in 1911 and saved as many of his lab notes as she could find. A few of those notes showed that Hesse and Lina had originated the idea of agar in microbial growth media, and they have since been recognized for their part in microbiology.

Three years after Koch and Hesse switched to agar-based media, another assistant in the laboratory, Richard J. Petri, designed a shallow glass dish to ease the dispensing of the sterilized molten media. The dishes measured a little less than a half-inch deep and 4 inches in diameter. This Petri dish design has never been improved upon and is a staple of every microbiology lab today.

Figure 1.3 Pouring molten agar. Agar melts when sterilized, and then solidifies when it cools to below 110°F. The microbiologist here pours the agar aseptically from a sterile bottle to a sterile Petri dish. (Courtesy of BioVir Laboratories, Inc.)

## The size of life

Bacteria need only be big enough to hold their vital enzymes, proteins, and genetic machinery. Evolution has eliminated all extraneous structures. Also, a small, simple architecture allows for rapid reproduction, which aids adaptation. Bacterial metabolism is a model of efficiency because of a large surface-to-volume ratio that smallness creates. No part of a bacterial cell is very far from the surface where nutrients enter and toxic wastes exit. Eukaryotic cells that make up humans, algae, redwoods, and protozoa contain varied organelles each surrounded by a membrane. The surface-to-volume ratio in these cells is one-tenth that of bacteria, so shuttling substances across all those organelle membranes, the cytoplasm, and the outer membrane burns energy. Bacterial structure is less demanding and more efficient. Finally, small size contributes to massive bacterial populations that dwarf the populations of any other biota.

Large multicellular beings that produce small litters with long life spans—think whales, elephants, and humans—take a long time to make new, favorable traits part of their genome. Insects evolve faster and can develop a new trait within a few years. In bacteria, evolution occurs overnight. Often, the progeny contain a new trait that makes them better equipped for survival.

No one knows the number of bacterial species. About 5,000 species have been characterized and another 10,000 have been partially identified. Biodiversity authority Edward O. Wilson has estimated that biology has identified no more than 10 percent of all species and possibly as little as 1 percent. Wilson's reasoning would put the total number of bacterial species at 100,000, probably a tenfold underestimate. Most environmental microbiologists believe that less than one-tenth of 1 percent of all bacteria can currently be grown in laboratories so that they can be identified.

Microbial geneticist J. Craig Venter's studies on microbial diversity have correctly pointed out that the number of species may be less important than their diversity and roles in the Earth's biosphere. Venter concluded from a two-year study of marine microbes that for every 200 miles of ocean, 85 percent of the species, judged by unique genetic sequences, changed. The ocean appears to contain millions of subenvironments rather than one massive marine environment, and each milliliter holds millions of bacteria. The actual number of bacteria in the oceans alone may exceed any previous estimates for the entire planet. In future studies of Earth's microbial ecology, the absolute number of species will probably never be determined.

Microbiologists begin defining the microbial world by taking samples from the environment and determining the types of bacteria found there. One of the first questions to answer is: Are any of these bacteria new, previously undiscovered species? To answer this, microbiologists must understand the species that have already been characterized, named, and accepted in biology, such as *E. coli*.

Taxonomists assign all living things to genus and species according to outward characteristics and the genetics of an organism. Until the late 1970s, microbiologists identified bacteria through enzyme activities, end products, nutrient needs, and appearance in a microscope. In 1977 Carl Woese at the University of Illinois proposed using fragments of a component of cell protein synthesis, ribosomal ribonucleic acid (rRNA). Cellular rRNA takes information contained in genes and helps convert this information into proteins of specific structure and function. Because the genetic information in rRNA is unique to each species, it can act as a type of bacterial fingerprint. Woese's method specifically used a component called 16S rRNA, which relates to a portion of the ribosome, the 16S subunit. This analysis led to a new hierarchy of living things (causing considerable consternation among traditional taxonomists) with bacteria, archaea, and eukaryotes comprising the three domains shown in Figure 1.4. Prior to the new rRNA classifications, biology students had been taught five-, six-, and even eight-kingdom classifications for organizing all plants, animals, and microbes. When I took my first biology classes, the five-kingdom system being taught looked like this:

- Monera, containing the bacteria
- Protista, containing protozoa and algae
- Plantae, containing green plants descended from algae
- Fungi descended from specific members of the Protista
- Animalia descended from specific members of the Protista

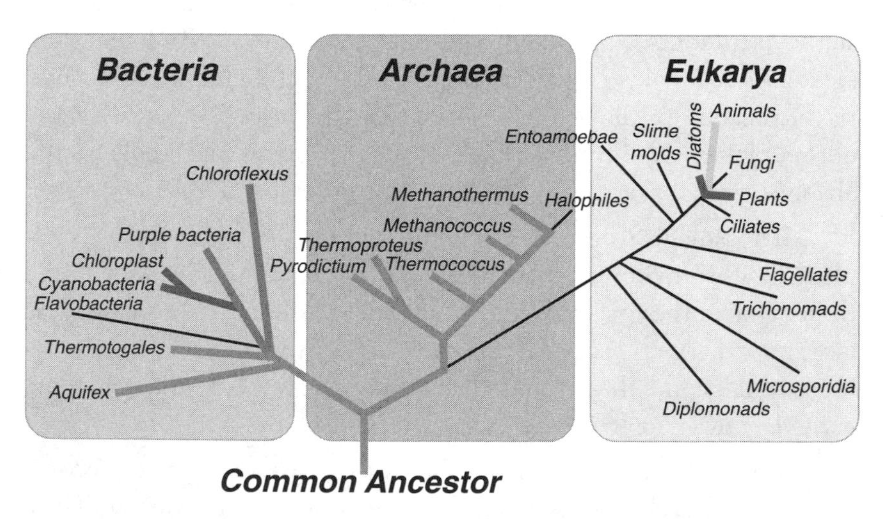


Figure 1.4 The three domains. Classification of the world's organisms does not remain static; new technologies constantly force taxonomists to reevaluate and reclassify species.

New technologies for classifying organisms have yet to end confusion that ensues when attempting to organize the world's biota, and for good reason. Taxonomists and philosophers have been trying to figure out organisms' relationships to each other since Aristotle's first attempts. Additionally, since the emergence of DNA analysis in the 1970s, geneticists have discovered more diversity in biota but also a dizzying amount of shared genes, especially among bacteria. The rRNA analysis introduced by Woese showed the degree to which different species shared genes. The studies revealed a significant amount of horizontal gene transfer, which is the appearance of common genes across many unrelated species.

The evolutionary tree we all learned in which families, genera, and species branched from a major trunk does not depict horizontal gene transfer. The evolutionary tree may look more like a bird's nest than an oak. Nowhere may that be truer than in the bacteria. Gene sharing or gene transfer is now known to take place in bacteria, and possibly archaea, more than ever before imagined. In 2002, the 16S rRNA system became further refined by focusing on certain protein-associated genes. But as biologists dig deeper into the genetic makeup of bacteria, they find more shared genes. Some microbiologists have begun to think that the term "species" makes no sense when speaking about bacteria. Currently, if two different strains of bacteria have less than 97 percent identical genes determined by 16S rRNA analysis, then they can be considered two different species. Some microbiologists suggest that only a 1 percent difference in genes differentiates species, not 3 percent.

When microbiologists first developed the bacterial groups known today as species, they let common characteristics of bacteria guide them. Gram reaction, nutrient requirements, unique enzymes, or motility served as features for putting bacteria into various species. Modern nucleic acid analysis has shown whether the traditional classification system still makes sense. With a high percentage of shared genes among bacteria and the ease with which diverse cells transfer genes around, some microbiologists have suggested that classifying bacteria by species is futile. It seems as if all bacteria belong to one mega-species, and different strains within this species differ by the genes they express and the genes they repress. By classifying bacteria into a single species, all bacteria would obey the definition for a species first proposed by Ernst Mayr in 1942: Members of the same species interbreed and members of different species do not.

Genetic analysis has blurred the lines between bacterial species so that the criteria used to classify other living things cannot apply to bacteria. To preserve their sanity, microbiologists need some sort of taxonomic organization so that they can speak the same language when discussing microbes. The traditional methods of grouping bacteria according to similar characteristics have turned out to be the handiest method regardless of DNA results. Microbiologists use the same classification and naming system for bacteria as used for all other life. The system has changed little since botanists in the mid-1800s, Carl Linnaeus being the most famous, developed it. Species classification and naming uses binomial nomenclature to identify every species by a unique two-part Latin name.

Bacteria of the same genus share certain genes, quite a few as mentioned, but different species have a few unique genes. For example, *Bacillus* is the genus name of a common soil bacterium. The genus contains several different species: *Bacillus subtilis* (shortened to *B. subtilis*), *B. anthracis*, *B. megaterium*, and so on. If I were a bacterium, my name would be *Maczulak anne* or *M. anne*.

To name a new bacterium, microbiologists have several conventions at their disposal. All that matters is that the new name be different from all other names in biology. Table 1.1 shows common naming conventions.

**Table 1.1** Origins of bacteria names

| Naming Method | Example | Reason for the Name |
|---|---|---|
| A historic event | *Legionella pneumophila* | Cause of a new disease that occurred at a Legionnaires convention in 1976 |
| Color | *Cyanobacterium* | Blue-green color |
| Cell shape and arrangement | *Streptococcus pyogenes* | Long, twisting chains (strepto-) of spherical (-coccus) cells |
| Place of discovery | *Thiomargarita namibiensis* | Found off the coast of Namibia |
| Discoverer | *Escherichia coli* | Discovered by Theodor Escherich in 1885 |

**Table 1.1** Origins of bacteria names

| Naming Method | Example | Reason for the Name |
|---|---|---|
| In honor of a famous microbiologist | *Pasteurella multocida* | Genus named for Louis Pasteur |
| Unique feature | *Magnetospirillum magnetotacticum* | Spiral-shaped bacteria with magnet-containing magnetosomes inside their cells |
| Extreme growing conditions | *Thermus aquaticus* | Grows in very hot waters such as hot springs |

Bacterial names will likely never be replaced regardless of scientific advances in classifying and reclassifying the species. Medicine, environmental science, food quality, manufacturing, and biotechnology depend on knowing the identity of a species that causes disease or contamination or makes a useful product. As microbiology fine-tunes its focus from the biosphere to the human body, species identity becomes more important.

## The bacteria of the human body

Ten trillion cells make up the human body, but more than ten times that many bacteria inhabit the skin, respiratory tract, mouth, and intestines. Microbiologists are fond of pointing out that if all of a person's DNA were mixed with the body's entire bacterial DNA, that person would be genetically more bacterial than human.

About 1,000 different species belonging to 200 genera live on the body rather than in it. An animal's body is a tube. The skin comprises the tube's outer surface, and the gastrointestinal tract from mouth to anus makes up the inner surface. The body's interior of blood, lymph, and organs normally contain no bacteria; these places are sterile. Urine and sweat exit the body as sterile fluids. In plants by contrast, bacteria live on but also inside the plant body.

The skin holds habitats that vary in moisture, oils, salts, and aeration. The scalp, face, chest and back, limbs, underarms, genitals, and feet make up the skin's main habitats, and each of these contains smaller, distinct living spaces. The entire skin surface has about one million bacteria on each square centimeter ($cm^2$) distributed unevenly among the habitats; the dry forearms contain about 1,000 bacteria per $cm^2$, and the underarms have many millions per $cm^2$.

Microbiologists sample skin bacteria by pressing a cylinder about the size of a shot glass open at both ends against the skin to form a cup, and then pouring in a small volume of water. By agitating the liquid and gently scraping the skin with a sterile plastic stick the microbiologist dislodges many of the bacteria. But no method or the strongest antiseptics remove all bacteria from the skin: The skin is not sterile. *Staphylococcus*, *Propionibacterium*, *Bacillus*, *Streptococcus*, *Corynebacterium*, *Neisseria*, and *Pseudomonas* dominate the skin flora.

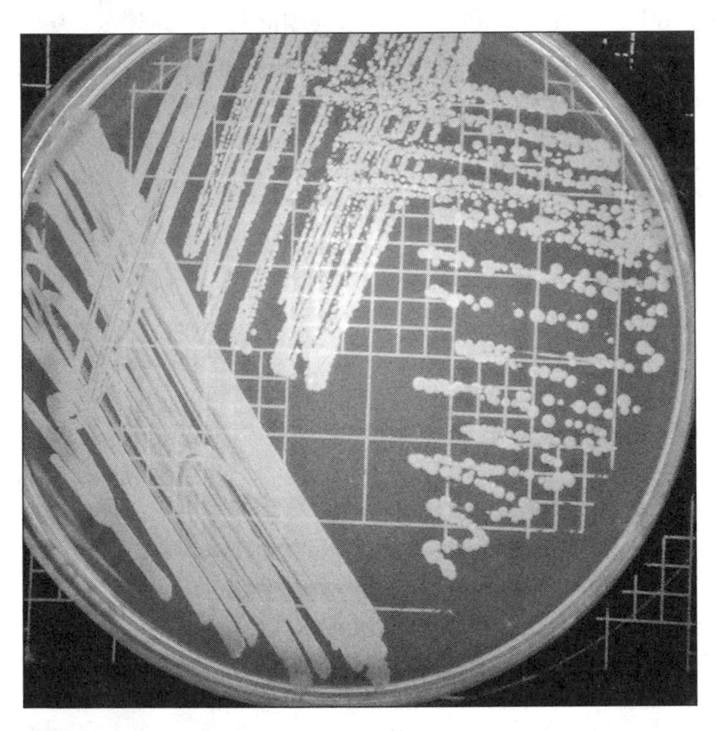

Figure 1.5 *Staphylococcus aureus*. A common and usually harmless inhabitant of skin, *S. aureus* can turn dangerous given the opportunity. This species can infect injuries to the skin, and the MRSA strain has become a significant antibiotic-resistant health risk. (Courtesy of BioVir Laboratories, Inc.)

Some of these names are familiar because they also cause illness, and yet a person's normal bacteria pose no problem on healthy, unbroken skin. The native flora in fact keep in check a variety of transient bacteria collected over the course of a day. Some of these transients might be pathogenic, but they do not settle permanently on the skin because the natives set up squatters' rights by dominating space and nutrients, and producing compounds—antibiotics and similar compounds called bacteriocins—that ward off intruders. Such silent battles occur continually and without a person's knowledge. Only when the protective barrier breaks due to a cut, scrape, or burn does infection gain an upper hand. Even harmless native flora can turn into opportunists and cause infection because conditions change in the body. Immune systems weakened by chemotherapy, organ transplant, or chronic disease increase the risk of these opportunistic infections:

- **Staphylococcus**—Wound infection
- **Propionibacterium**—Acne
- **Bacillus**—Foodborne illness
- **Streptococcus**—Sore throat
- **Corynebacterium**—Endocarditis
- **Pseudomonas**—Burn infection

Anaerobic bacteria do not survive in the presence of oxygen, but they make up a large proportion of skin flora. Though the skin receives a constant bathing of air, anaerobes thrive in miniscule places called microhabitats where oxygen is scarce. Chapped and flakey skin and minor cuts create anaerobic microhabitats. Necrotic tissue associated with major wounds also attracts anaerobes, explaining why gangrene (caused by the anaerobe *Clostridium perfringens*) and tetanus (*C. tetani*) can develop in improperly tended injuries. Of normal anaerobes inhabiting the skin, *Propionibacterium acnes* (the cause of skin acne), *Corynebacterium*, *Peptostreptococcus*, *Bacteroides*, and additional *Clostridium* dominate.

The mouth's supply of nutrients, water, and microhabitats creates a rich bacterial community. Brushing and flossing remove most but not all food from between teeth, the periodontal pockets between the tooth and the gum, and plaque biofilm on the tooth surface, which holds a mixture of proteins, human cells, and bacterial cells. Anaerobes and aerobes find these places and their relative numbers vary from daytime to night as the level of aeration, flushing with drinks, and saliva production changes. During the day, more air bathes oral surfaces and aerobes flourish. At night or during long periods of fasting, the aerobes consume oxygen and anaerobes begin to prosper. By the nature of their fermentations, anaerobes make malodorous end products when they digest food. These bad-smelling, sulfur-containing molecules vaporize into the air and become bad breath.

Few bacteria live in the esophagus and stomach with the exception of the spiral-shaped *Helicobacter pylori*, occurring in half of all people with peptic ulcers. The discovery of *H. pylori* in the stomach in 1975 dispelled the long-held belief that no microorganisms could withstand the digestive enzymes and hydrochloric acid in gastric juice. Most bacteria traverse the half gallon of stomach fluid at pH 2 by hiding in a protective coat of food particles on the way to the small intestine. *H. pylori*, however, thrives in the stomach by burrowing into the mucus that coats the stomach and protects the organ from its own acids. Inside the mucus, the bacteria secrete the enzyme urease that cleaves urea in saliva into carbonate and ammonia. Both compounds create an alkaline shield around *H. pylori* cells that neutralize the acids.

The pH rises in the intestines and bacterial numbers increase a millionfold from about 1,000 cells per gram of stomach contents, which to a microbiologist is a small number. Humans, cows, pigs, termites, cockroaches, and almost every other animal rely on intestinal bacteria to participate in the enzymatic digestion of food. The numbers reach $10^{12}$ cells per gram of digested material. Monogastric animals such as humans and swine absorb nutrients made available by the body's enzymes as well as nutrients produced by bacteria. When the bacteria die and disintegrate in the intestines, the body absorbs the bacterial sugars, amino acids, and vitamins (B-complex and vitamin K) the same as dietary nutrients are absorbed. Cattle, goats, rabbits, horses, cockroaches, and termites, by contrast, eat a fibrous diet high in cellulose and lignin that their bacteria must break down into compounds called volatile fatty acids. Glucose serves as the main energy compound for humans, but volatile fatty acids power ruminant animals (cattle, sheep and goats, elephants, and giraffes) and animals with an active cecum (horses and rabbits).

Rumen bacteria carry out anaerobic fermentations. Almost every organic compound in the rumen becomes saturated there by fermentative bacteria before moving on to the intestines. As a result, ruminants such as beef cattle deposit saturated fats in their body tissue. Nonruminant animals, such as pigs and chicken, carry out fermentations to a lesser extent and their meat contains less saturated fat.

How important are all these bacteria in keeping animals alive? Germfree guinea pigs grow smaller than normal, develop poor hair coat, and show symptoms of vitamin deficiency compared with animals with a normal microbial population. Germfree animals also catch infections more than populated animals. On the upside, germfree animals never experience tooth decay!

*Bacteroides*, *Eubacterium*, *Peptostreptococcus*, *Bifidobacterium*, *Fusobacterium*, *Streptococcus*, *Lactobacillus*, and *E. coli* of the human intestines also produce heat in the same way wine fermentations produce heat. This heat loss is inefficient for the bacteria—any energy that dissipates before it can be used is lost forever—but the body uses it to maintain body temperature. The large numbers of normal intestinal bacteria also outcompete small doses of food illness bacteria such as *Salmonella*, *Clostridium*, *Bacillus*, *Campylobacter*, *Shigella*, *Listeria*, and *E. coli*.

*E. coli* is the most notorious of foodborne pathogens and also the most studied organism in biology. In fact, *E. coli* plays a minor role in the digestive tract; other bacteria outnumber it by almost 1,000 to one. *E. coli* has become the number one research tool in microbiology for two reasons. First, this microbe cooperates in the laboratory. *E. coli* is a facultative anaerobe, meaning it grows as well with oxygen as without it. It requires no exotic nutrients or incubation conditions, and it doubles in number so rapidly that a microbiologist can inoculate it into nutrient broth in the morning and have many millions of cells that afternoon. The second reason for using *E. coli* in biology relates to the ease of finding it in nature: The human bowel and that of most other mammals produce a constant supply.

## The origins of our bacteria

Infants have no bacteria at birth but start establishing their skin flora within minutes and digestive tract populations soon after. *E. coli*, lactobacilli, and intestinal cocci latch on to a baby during birth and become the first colonizers of the infant's digestive tract. Babies get additional bacteria for a reason that scares germophobes: fecal and nonfecal bacteria are everywhere, and people ingest large amounts each day. Fecal bacteria disseminate beyond the bathroom to countertops, desks, refrigerator handles, keyboards, remote controls, and copy machine buttons. Any object repeatedly touched by different people contains fecal bacteria. Newborns get these bacteria every time they handle toys or crawl on the floor, and then put their hands or other objects in their mouth. Adults similarly receive fecal bacteria, called self-inoculation, when touching their hands to the mouth, eyes, or nose. Adults touch their hands to their face hundreds of times a day, and children do it more frequently.

A baby's digestive tract has some oxygen in it so aerobic bacteria and facultative anaerobes prosper there first. *E. coli* colonizes the gut early on and uses up the oxygen. A population of anaerobes then begins to dominate: *Bacteroides*, *Bifidobacterium*, *Enterococcus*, and *Streptococcus* make up the common genera. The adult digestive tract distal to the mouth will eventually contain 500 to 1,000 different species of bacteria and a lesser number of protozoa.

Pathogens make up a minority of all bacteria, but the word "germs" brings only bad connotations. A growing number of microbiologists have nonetheless begun to see the potential benefits of exposure to germs. In the 1980s German pediatrician Erika von Mutius investigated the apparent high incidences of asthma and allergies in industrialized nations compared with developing areas. She compared the health of children from households that received little housekeeping with counterparts in well-managed households with regular cleanings. Children who had been exposed to a dirty environment had fewer respiratory problems than children from cleaner surroundings. Von Mutius therefore proposed that a steady exposure to germs might help youngsters develop strong immune systems.

Von Mutius's "hygiene hypothesis" drew criticism from microbiologists and, unsurprisingly, manufacturers of cleaning products. But pediatric allergist Marc McMorris supported the hypothesis, saying, "The natural immune system does not have as much to do as it did 50 years ago because we've increased our efforts to protect our children from dirt and germs."

Questions have not yet been answered on whether continued use of disinfectants and antimicrobial soaps change bacteria at the gene level. Medical microbiologist Stuart Levy has argued that antibiotic overuse combined with overzealous use of antimicrobials leads to bacteria resistant to the chemicals meant to kill them. These bacteria may develop subsequent resistance to antibiotics. Bacteria eject harmful chemicals and also antibiotics from inside the cell by using a pumplike mechanism. If bacteria use the very same pump for chemical disinfectants as for antibiotics, the vision of a new generation of super-resistant bacteria becomes probable. Imagine hospitals where no antibiotics can stop pathogens and few chemical disinfectants can kill them. Doctors and microbiologists have warned that medicine is inching closer to this very scenario.

The body helps native flora defend against pathogens that attach to the skin. The enzyme lysozyme in tears and saliva kills bacteria, and skin oils contain fatty acids that inhibit gram-positive bacteria. If those defenses fail, the immune system sets in motion a hierarchy of defenses meant to find and destroy any foreign matter in the bloodstream.

Dental caries can lead to more serious tooth decay and gum disease, or an infection of the blood if the oral lesions are severe. In plaque, *Streptococcus mutans*, *S. sobrinus*, and various lactobacilli (lactic acid-producing bacteria) initiate caries formation by producing acids. Lactic, acetic (also in vinegar), propionic, and formic acid diffuse into the tooth enamel and break it down by demineralization, meaning the removal of minerals such as calcium. Demineralization occurs several times a day in a cycle in which new dietary calcium and phosphate and fluoride from toothpaste replace the lost minerals. Dental caries offer an exception to the rule that native flora do not initiate infection.

On the skin, some bacteria create a nuisance. Skin bacteria consume amino acids, salts, and water excreted by eccrine sweat glands. These glands located all over the body produce copious amounts of watery sweat for cooling. The bacteria also feed on thicker sweat from apocrine glands in the underarms, ear canal, breasts, and external genitalia. These glands tend to activate in times of stress or sexual stimulation. Skin bacteria in these places degrade the sweat's sebaceous oils to a mixture of small fatty acids and nitrogen- and sulfur-containing compounds, all of which vaporize into the air to cause body odor.

Some bacteria such as *Staphylococcus* live on everyone, but each person also has a unique population of native bacteria that produces a distinctive odor. Scientists have long sought elusive secretions called pheromones that foster communication between people through smell, but I suspect the secretions of native flora will prove to be the human version of quorum sensing. In 2009 anthropologist Stefano Vaglio analyzed the volatile compounds in the sweat of women shortly after childbirth and discovered unique patterns of odor compounds, perhaps to aid mother-infant recognition.

The deodorant and soap industries spend a fortune convincing people to block the natural products made by skin bacteria. Each week hundreds of deodorant-testing volunteers troop into deodorant companies' odor rooms. The volunteers take positions like a police lineup and raise their arms. A team of trained sniffers works its way down the line to "score" the results. Women make up the majority of professional sniffers; the Monell Chemical Sciences Center confirmed in 2009 that women's olfactory systems gather more information from body odors than men's. (Sniffers have sworn that if blindfolded they could identify their mates.) The sniffers assess the best and the worst new deodorants based on underarm odor scores; 0 equals no odor and a score of 10 could clear a room.

## One planet

During the Golden Age of Microbiology, bacteria were viewed as unrelated individualists. Pasteur studied the bacteria that made lactic acid by fermenting sugar. Joseph Lister focused on germs causing infections in hospital patients. Robert Koch discovered the anthrax pathogen, *Bacillus anthracis*, and delved into the processes of bacterial disease. He would develop a set of criteria (Koch's postulates) that gave birth to today's methods for diagnosing infectious disease. Not until microbial ecology developed did biologists recognize the interrelated world of bacteria as well as the relationship between environmental bacteria and humans.

*Staphylococcus epidermidis* contributes to body odor, a bacteria-human connection easily detected. But thousands of hidden bacterial activities shape the very ecology of the planet. In soil, *Azotobacter* pulls nitrogen from the air, chemically rearranges it, and hands it off to *Nitrosomonas*, which changes the nitrogen again and shuttles it to *Nitrobacter*. *Nitrobacter* then secretes the nitrogen in the form of nitrate, which disseminates throughout soils. Some of the nitrate reaches the roots of legumes such as clover or soybeans. Inside the plant roots anaerobic *Rhizobium* absorbs the nitrate and converts it to a form the plant can use. This process is vital in replenishing nitrogen that higher organisms need.

For carbon to make a similar cycle through the Earth's organic and inorganic matter, the bacteria of decay must help decompose the planet's fallen trees, plants, and animals. The common soil inhabitant *Bacillus* breaks down proteins, fats, and carbohydrates by excreting the enzymes protease, lipase, and amylase, respectively. Thousands of other species break down organic matter in similar ways. For example, *Cellulomonas* bacteria produce the enzyme cellulase—rare for bacteria—that digests plant cellulose fibers. Bacteria emit carbon dioxide as an end product, which enters the atmosphere. A massive population of photosynthetic bacteria in the Earth's surface waters then captures this gas and inserts the carbon into a new food chain of bacterial cells, protozoa, invertebrates, and so on until the carbon ends up in tuna sashimi on a restaurant menu.

If clouds begin to form while a person lunches on sashimi, bacteria have a part in that, too. Photosynthetic marine bacteria and algae produce dimethyl sulfide gas as a waste product of their normal metabolism; they emit 50 million tons annually. When the gas rises and enters the atmosphere, it chemically rearranges into sulfate, which attracts water vapor. The vapor turns to droplets and forms clouds. On a global scale clouds inhibit the photosynthetic bacteria and less dimethyl sulfide forms. When the clouds thin, the cycle begins again.

Albert Kluyver of the Technical School of Delft—the town where van Leeuwenhoek discovered bacteria in 1677—praised the wonderful "unity and diversity" of microorganisms, a perfect description for dissimilar organisms that share more than 95 percent of their genes. The human body possesses its own unity and diversity of microbes that in most situations keep the body's metabolism working at its best. Pathogens more than good bacteria gain the attention of researchers and doctors. For this reason, epidemics have expanded our knowledge of bacteria. Many of the discoveries in microbiology came about from a blend of genius and serendipity, a fair description of all science.

# 2

## Bacteria in history

Other than infectious disease, humanity's early dealings with bacteria involved mainly the production of foods. Wheaton College biologist Betsey Dexter Dyer once noted that a meal can be assembled completely from bacteria-produced foods, such as the following items.

- **Cheeses**—Swiss from *Propionibacterium* and limburger from *Brevibacterium*
- **Olives**—*Leuconostoc*, *Lactobacillus*, and *Pediococcus*
- **Dry sausages**—*Pediococcus*
- **Sourdough bread**—Various lactic acid-producing bacteria
- **Butter**—*Lactobacillus*
- **Cottage cheese**—*Streptococcus*

A steak or a glass of milk results from the digestion of grasses by anaerobic bacteria in the rumen of cattle. The rumen fermentations are oxygen-free conversions of sugar into microbial energy with acid or alcohol as a by-product.

Olives may be the oldest food fermented specifically to make a new food. The Phoenicians brought olives throughout the Greek isles by 1600 BCE. The production of acids in the fermentation process helped preserve the product during long sea voyages. No one knows who made this discovery, but food historians assume that fermented foods were discovered by accident or perhaps by necessity by explorers who had already eaten all other food supplies.

Bacterial food spoilage takes the form of acid production, protein curdling, gas or toxin production, or decomposition. The latter two cause foodborne illness or a loss of food's nutritional value, respectively, and render the food unusable. Properly controlled acid production, however, preserves fresh vegetables, fruits, and juices and retains most nutrients, while protein curdling does the same in dairy products.

Evidence of winemaking from alcohol-producing bacteria dates to 6000 BCE Mesopotamia and no doubt started earlier. Over the next two millennia, Hebrew, Chinese, and Inca cultures perfected yeast fermentations for wines and beers, but retained bacteria for fermenting crops to make sauerkraut, pickles, wine, soy sauce, silage, and other foods that lasted longer with an acid preservative than in the fresh form. The names of brave souls who tasted spoiled foods have been lost to history, but either by necessity or a sense of adventure, they invented food preservation.

Bacteria-made dairy products date to before 3000 BCE using milk from cows, yaks, goats, sheep, horses, camels, and even reindeer. Fermented milk products, the "mere white curd of ass's milk" as described by 18th century poet Alexander Pope likely originated in more than one place. Traders used pouches made of cleaned animal entrails for carrying milk between villages and would not have realized that the stomach enzyme rennin (also called chymosin) remained active in the pouch lining. This enzyme helps nursing infants digest milk by curdling the milk proteins and thus slowing their passage through the digestive tract. In a pouch slung over a horse's rump, the rennin made cheese.

Lactic acid-producing *Lactobacillus*, *Lactococcus*, *Streptococcus*, and *Leuconostoc* make up the main bacteria used in cheeses, yogurt, butter, buttermilk, and sour cream today as they did centuries ago. Manufacturers of salad dressings, coleslaw mixes, and mayonnaise now encourage the growth of lactic acid bacteria to produce an acidic tangy flavor and preserve the food.

When bacterial contaminants did not produce a tasty, edible product, the ancients froze, smoked, or dried the food or added salt, sugar, or honey. These preservation methods inhibit bacteria's growth by making water molecules unavailable for cellular reactions. Food producers still use these ancient methods, but they now also use chemicals to inhibit the growth of microbes in food.

Bacteria have ploys for escaping physical injury from lack of water or harm from chemicals. Many bacteria enter a state of dormancy when water becomes scarce and grow again when water returns to their environment. The normal soil inhabitants *Clostridium* and *Bacillus* have evolved an adaptation that protects better than dormancy: the formation of endospores. More than any other type of cell in biology, endospores resist freezing, heating, boiling, chemicals, and irradiation. A microbiologist need only dilute a small amount of soil in nutrient broth and then incubate it to make the endospores germinate into actively growing cells. (Sometimes stubborn endospores need to be heat-shocked at 130°F for five minutes before they will germinate.)

In 1993, American microbiologists Raúl Cano and Monica Borucki found endospores resembling *Bacillus sphaericus* in an extinct bee that had been preserved in amber estimated at 25 to 40 million years old. As is customary in science when radically new discoveries are made, skeptics came forward suggesting the bacteria were contaminants from a later period. The critics charged that no living organism can survive that long. But in 2000, biologist Russell Vreeland found *Bacillus* endospores buried in 250-million-year-old salt deposits and showed they remained viable by growing the cells in his laboratory. Vreeland and his team then completed 16S rRNA analysis on the microbe and identified it as an ancestor of modern *Bacillus*. Perhaps expecting the same skepticism Cano had met with, Vreeland also calculated the chances of a contaminant invading the sterilized equipment or breeching his aseptic techniques at one in one billion. Assuming these bacteria are not contaminants, research like this demonstrates the astonishing durability of bacterial endospores and also hints at the challenges of protecting food from spore-forming pathogens.

## The ancients

Paleopathology is the investigation of ancient artifacts for clues on history's diseases (see Figure 2.1). Paleopathologists use fiber optics, X-ray imaging, and computerized tomography to see inside caskets without disturbing the contents. Only when they find evidence of damaged tissue do they open the casket and salvage DNA from a bit of tissue, bone, or tooth pulp. By comparing the ancient DNA with that of present-day pathogens, scientists have identified the main bacterial diseases that have haunted society for millennia: anthrax, bubonic plague (*Yersinia pestis*), cholera (*Vibrio cholera*), diphtheria (*Corynebacterium diphtheria*), leprosy (*Mycobacterium leprae*), syphilis (*Treponema pallidum*), tuberculosis (TB) (*M. tuberculosis*), and typhoid fever (*Salmonella typhi*). Facts gleaned from ancient writings have supplemented the technology of paleopathology. Pliny the Younger wrote of Roman society from 79 to 109 CE and in one essay described an illness affecting a close friend:

> She has continued fever, her cough gets worse day by day, she is very thin and weak. Still she is mentally alert, and her spirit does not flag, a spirit worthy of her husband Helvidius....In everything else she is failing to such an extent that I not only fear but grieve.

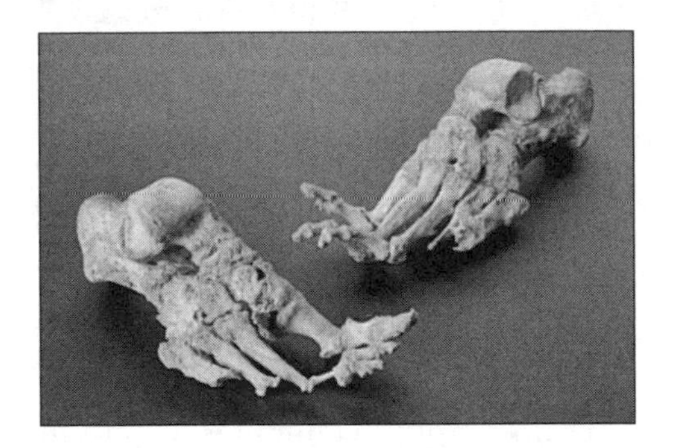

Figure 2.1 Leprosy. *Mycobacterium leprae* preferentially attacks the cooler extremities of the body, mainly skin and peripheral nerves. The disease erodes the skeleton, such as these feet dated to a leprosy sufferer from c. 1350. (Courtesy of Science and Society Picture Library, Science Museum, London)

The mention of coughing and weakness without reference to fever or delirium suggested to medical historians that Pliny wrote of tuberculosis. Studies on the emergence of diseases have been aided by the knowledge that cancer and heart disease were rare in antiquity; most deaths from disease can be attributed to infectious diseases.

Some people sensed that hygiene affected quality of life 1,000 years before microbiologists connected bacteria with disease. Mesopotamia's Sargon I decreed the construction of privies for the ruling class in the 3rd century BCE, and the Greeks and Egyptians devised similar toiletlike receptacles to protect drinking water and food from human waste. The Roman Empire's largest cities established a model for sanitation infrastructure with freshwater aqueducts, public baths, and sewers for the wealthy. (Rome's poor endured squalid conditions that led to chronic infections and short lives.) Romans sprinkled spices and herb oils into bath waters for fragrance. These substances are now known to kill bacteria when used in low concentrations.

Hygiene practices changed when the Roman Empire declined. The Roman Catholic Church took a bigger role in influencing public opinion as well as science, teaching that disease came from God as punishment for evil; some present-day clergy continue to embrace this belief. Human behavior certainly influences disease transmission, but evil has nothing to do with it.

## The legacy of bacterial pathogens

During World War II, scientists in Germany and Great Britain raced to find a "magic bullet." They sought not a weapon but a drug to stop needless fatalities from infections of battlefield injuries. Before the magic bullet arrived, herbs served as the main way to fight infectious disease, with mixed results.

Evidence of TB predates written historical records. *Mycobacterium bovis* may have entered the human population between 8000 and 4000 BCE with the domestication of cattle. Samples from the spinal column of Egyptian mummies from 3700 BCE have shown signs of damage from the disease, but no one has determined if the infection had come from *M. bovis* or the cause of modern TB, *M. tuberculosis*. The distinction is small because these two species share more than 99.5 percent of their genes.

In 400 BCE Hippocrates identified the most widespread disease in Greece as phthisis and in so doing described classic signs of consumption or TB. In *Aphorisms*, he warned, "In persons affected with phthisis, if the sputa which they cough up have a heavy smell when poured upon coals, and if the hairs of the head fall off, the case will prove fatal." The infected transmit the TB pathogen when they cough, sneeze, or merely breathe, and dense population centers have always acted as breeding grounds. Today TB infects one-third of the world's population, and humans, not cattle, serve as the reservoir.

Bubonic plague, syphilis, and anthrax began appearing with regularity about 2000 BCE. Historians have debated the message intended by historian Ipuwer, who sometime between 1640 and 1550 BCE wrote of the Egyptian plagues. In referring to the Fifth Egyptian Plague he wrote, "All animals, their hearts weep. Cattle moan." This passage seems to describe anthrax, a disease as deadly in animals as in humans.

Mobile societies enabled infectious diseases to reach wider distribution. Infectious agents traveled with Egyptian and Phoenician traders crisscrossing the Mediterranean between 2800 and 300 BCE. Both cultures sent ships into the Red Sea and to Persia, but the Phoenicians also sailed north along Europe's coast. If local residents managed to evade syphilis-infected sailors, they might have contracted disease from some of the tradable goods infested with parasites and pathogens. Anthrax endospores, for instance, hid in hides, pelts, and wool that held bits of soil. As each ship docked, rodents undoubtedly clambered down gangways and brought bubonic plague to shore.

Anthrax became a disease of laborers. Anyone who worked their hands into the soil had a much greater chance of inhaling *B. anthracis* endospores or infecting a cut. Shearers, tanners, and butchers also had higher incidences of the disease than the rest of society. Since livestock also picked up the microbe from the ground, anthrax caused occasional epidemics in agriculture. The Black Bane of the 1600s killed nearly 100,000 cattle in Europe. People do not transmit anthrax to each other; infection comes mainly by inhaling endospores from the environment. When *B. anthracis* germinates in the lungs, mortality rates reach 75 percent of infected individuals. Today, the United States has less than one case of anthrax a year. Slightly higher rates occur in the developing world.

Syphilis-causing *Treponema* might also have entered the human population from animals, perhaps in tropical Africa. Bacteria similar to the one causing human syphilis—the Great Pox—were isolated in 1962 from a baboon in Guinea, but few other clues about the origin of syphilis exist. Ancient explorers brought syphilis throughout the

Mediterranean and Europe. Its migration mirrored the spread of the major bubonic plagues that followed trade routes west from Asia to Europe, and later followed the slave trade from Africa to the western hemisphere. Syphilis additionally accompanied each of history's armed invasions.

Disease historians diagnose syphilis in skeletons by looking for the presence of *caries sicca* (bone destruction) of the skull, which gives the bone a moth-eaten appearance plus characteristic thickening of the long bones. Corkscrew-shaped *T. palladium* wriggles into the testicles where it reproduces and then infects a sexual partner by similarly burrowing into the skin. The bacteria then enter the lymph system and bloodstream. As syphilis progresses, the skin, aorta, bones, and central nervous system are affected, but the disease's early-stage signs are so nebulous that misdiagnosis persisted for centuries. Physicians could not distinguish syphilis from leprosy until Europe's first major syphilis outbreak from 1493 to 1495 in Naples, which remains one of history's worst syphilis epidemics. The siege of Naples has also been implicated as the origin of syphilis in the New World, a debate that continues to this day.

In 1493, France's Charles VIII claimed Naples as his by birthright and sent his army to wrest it from Spain. During the clash, syphilis spread from Naples to the rest of Europe. The timing of the siege and Christopher Columbus's voyages from Spain have convinced some historians that Columbus's men brought syphilis to the Americas. Columbus left the port of Palos, Spain, with three ships and 150 men in August, 1492, and returned in March, 1493, leaving dozens of his crew on the island of Hispaniola. The next two excursions from Cadiz to Hispaniola totaled 30 ships and at least 2,000 men with return crossings in 1494 and 1495. After each voyage, most of Columbus's men wanted no more part of the open ocean and earned money by joining ranks with the Neapolitan troops to fight the French. When Naples finally rebuffed the invasion in 1496, Charles's troops returned home and syphilis went with them.

Writings by European physicians from 1497 to 1500 indicate that they had never before seen the disease that had first victimized Naples. The French called syphilis "the disease of Naples," but the Italians felt equally sure of the source. In 1500, Spanish physician

Gaspar Torella wrote, "On this account it was christened the *morbus Gallicus* by the Italians, who thought it was a disease peculiar to the French nation." The argument would never be resolved and the finger pointing probably continued for years.

Did Christopher Columbus's ships bring syphilis to the Americas as many historians believe? Martín Alonso Pinzón captained the *Pinta* in Columbus's first voyage. A bitter rival of Columbus throughout their sailing careers, Pinzón died of syphilis in 1493 soon after returning home to Spain. Although symptoms begin earlier, the fatal stages of the disease can arise 10 to 20 years after infection, which suggests that Pinzón may have contracted syphilis well before 1492. Before sailing with Columbus, Pinzón had voyaged along the African coast and to the Azores, widening the possibilities of where he caught syphilis. Although speculation on whether Columbus brought venereal disease to the Americas persists, the establishment of settlers' colonies would have increased the chance for disease transmission. If not Columbus, then certainly others who followed brought with them contagious diseases.

## The plague

Justinian I, 6th-century ruler of the Byzantine Empire, devoted himself to spreading Byzantine architecture from his throne in Constantinople, along the Mediterranean rim, up the Nile, and deep into Europe. To prepare for his fleet's voyages, Justinian ordered continuous stocking of the massive granaries on the city's outskirts. The grain sustained the ships, but also fed an exploding rat population.

By 540 CE Justinian had succeeded in expanding Constantinople's influence. But at each new port, residents fell nauseous and developed chills, fever, and headache, some within only two days of a ship's arrival. Their abdomen would swell with pain and bloody diarrhea followed. Their lymph nodes (or buboes) clogged with necrotic tissue and by six days of the first discomfort, many had died, the skin covered with dark purple lesions. The same occurred in Constantinople where deaths grew to 10,000 daily. Many who felt the first symptoms of illness panicked and fled to the countryside. Within days the fatalities rose in those rural places, too.

Justinian blinded himself to the misery and more than one assassination attempt. He drained the coffers to expedite his dream and perhaps also to entice new sailors from a dwindling labor pool. The Plague of Justinian would kill 60 percent of the empire or 100 million people by the time it had run its course in 590 CE. Justinian himself avoided the plague and died of natural causes at age 38.

Did a mitigating factor suddenly appear to cause the first bubonic plague in recorded history to arise? Rodents then as now carry the intestinal bacterium *Y. pestis* that can contaminate the animal's fur or skin. Fleas ingest *Y. pestis* each time they bite and so engorge themselves that their digestive tract fills with bacteria. The insect must regurgitate some bacteria just to stay alive. When a flea upchucks on an uninfected rodent, it transmits *Y. pestis* and creates an ever-expanding reservoir of disease. Poor sanitation leading to large rat populations in metropolises like Constantinople increased the probability of receiving flea bites. Justinian gave the epidemic a boost by building granaries that all but guaranteed a massive rat colony to serve as the disease's reservoir.

Following Justinian's rule, bubonic plagues mysteriously disappeared for the next 700 years. As the last remnants of Roman influence faded, disease control also declined, and many contradictions took root. People believed that retaining wastes and even an animal carcass in the home repelled evil and thus disease, yet many also assumed that bad odors brought illness—a cadaver in the living room surely smells. Even with the building of the first centralized hospitals at the dawn of the Middle Ages, medicine remained the domain of healers who used leeches for extracting the body's pains. Faulty birthing methods caused a high incidence of mental illness that further threatened good personal hygiene.

Beginning in the 14th century, four plague epidemics would decimate Europe, none more brutal than the Black Death, named for the black-purplish lesions formed by hemorrhaged vessels under the skin. Between 1346 and 1352 the Black Death killed more than 25 million people in Europe or about 30 percent of the total population. Combined with the loss of life as the plague followed trade routes from Asia in the 13th century and to northern Africa and the Crimea before reaching Europe, the global Black Death killed a total of 200 million.

As in Justinian's day, survivors could not bury the dead fast enough. Survivors carried the corpses on long poles—"I wouldn't touch that with a 10-foot pole"—to mass graves outside the towns. The epidemic slowed only when it reached the Alps where colder weather repelled rats, and the pathogen had likely mutated to a less virulent form.

Figure 2.2 Dance of Death. Death became an everyday occurrence, by the hundreds in some towns, in the Middle Ages in Europe. Artists, writers, and composers depicted bleak futures where Death overwhelms the living.

An epidemic that destroys 100 million lives in less than a decade and reduces Europe's population by one-third, as the Black Death did, certainly impacts society in ways that are felt for generations. Even art and music reflected the looming presence of Death, which usually triumphed over mortals (see Figure 2.2). Some cities lost 75 percent of their children, and entire family trees had been reduced to one individual—the plague had created a parentless generation. Craftsmen, artists, farmers, and clergy disappeared. A plunge in economic vitality caused birth rates to drop.

During the plagues, clergymen insisted as they had for centuries that sickness came as penance from God. Their ineffective efforts to administer to the dying by combining faith and sorcery caused the church to lose its customary privileged status in society. The banking profession gained stature, however, for two reasons. The plague's survivors understood the need for protecting assets for the next generation, especially when death could strike so suddenly. At the same time, serfs abandoned fields controlled by feudal landowners and took advantage of monetary pay to fill labor shortages in the cities. This in turn helped create a mobile workforce that covered Europe in search of the highest wages in labor-starved towns. Young adults left as the sole managers of family property opted against traveling to traditional centers of learning in Paris, Vienna, or Bologna. New centers of education thus developed in Oxford, Cambridge, Edinburgh, Amsterdam, Copenhagen, and Stockholm by the 14th century. The depopulation of the European continent also opened up new land for cultivation or development and laid the foundation for the industrial centers of today's Europe.

Surgeons had been as useless as clergymen during the plagues. The status of the surgeon would decline and not rebound until the mid-19th century when Joseph Lister invoked the need for sterile conditions in hospitals. Barbers came to the fore as more trustworthy medical practitioners despite their penchant for bloodletting as an all-purpose cure. But what is now known as western medicine also advanced. Medical schools grew and students for the first time learned anatomy and physiology. As a result, the medical community began learning about the effects of infectious disease on internal organs.

With each of these historical plagues, survivors learned better precautions for escaping infection. Bubonic plague is not contagious, but streets filled with the dead and dying certainly showed that anyone could fall victim. Plague survivors gingerly removed the bodies and took them to the countryside where funeral pyres awaited. This had been the commonest method of disposal throughout the Middle Ages, but on occasion people used more imaginative ways to dispose of corpses.

From 1344 through 1347, Tartars laid siege repeatedly to the port city of Caffa (now Feodosija, Ukraine), home to diverse nationalities and political persuasions. The plague had already laid waste to the Tartars' homeland of eastern Asia, and deaths among them mounted even as they surrounded the city. With a body count mounting, the Tartars disposed of their deceased by the simple expediency of catapulting the cadavers over Caffa's walls. Caffa's healthy residents would be infected when they collected the bodies for burial. Thus bacteria and humans forged a complex relationship involving disease, sustenance, evil, and God.

## Microbiologists save the day

In 1822 Louis Pasteur was born into a family that had made its living tanning hides for generations. A lackluster student, only chemistry held Pasteur's interest. By the time he reached college, Pasteur would spend hours studying structures of organic compounds and this pursuit likely awakened a curiosity about biology. Still, Pasteur thought of himself as foremost a chemist.

After winning election as France's president in 1848, Napoleon Bonaparte III made transportation, architecture, and agriculture the country's priorities. New edicts pressured university scientists to follow commercial pursuits. As a professor at the University of Lille, Pasteur grudgingly tucked away his chemistry equipment and brought a microscope into his lab without a clear plan for using it. He decided to teach students about biology's relationship to agriculture until the time came when he could return to his chemistry experiments.

Pasteur's "temporary" foray into biology initiated the most accomplished career in microbiology's history. His publication list lengthened, and his reputation grew inside and outside of science. By the 1850s, Pasteur had been recruited by France's alcohol manufacturers to improve their fermentation methods. He began by investigating yeast fermentations, perhaps because brewers had not studied it in detail. Pasteur noticed that a drop of liquid from the fermentation flasks gave a curious result in the microscope. When Pasteur put a glass cover slip on top of the drop, some of the microbes avoided the edges of the slip where the liquid was exposed to the air. Pasteur introduced biology to anaerobic bacteria.

By describing processes taking place in fermentation, Pasteur gave the wine and brewing industries greater control over their manufacturing steps. His reputation soared when he diagnosed a disease that had been decimating France's silk industry. By the 1860s, Pasteur reached national hero status. (Pasteur had incorrectly identified bacteria as the cause of the silkworm disease. Electron microscopes were not yet available to enable him to find the real cause: a virus. Pasteur nevertheless made the crucial and previously overlooked connection between microbes and infection.)

The public adored Louis Pasteur. Napoleon III invited the microbiologist to his table to hear the latest theories on microbes, and

Pasteur happily obliged. He, in fact, had developed the habit of dismissing anyone who questioned his work. Pasteur also cultivated the ability (or flaw according to some) of drawing scientific conclusions even while producing little data to back them. Louis Pasteur possessed such a rare and keen insight into biology that his conclusions almost always proved to be correct. One famous misstep occurred in 1865 during a cholera outbreak in Paris. Pasteur believed the pathogen *Vibrio cholerae* transmitted through the air though it is a waterborne pathogen. The French nonetheless felt relieved to know that Pasteur was hard at work trying to save them from cholera. The Paris epidemic ran its course and disappeared on its own.

During a rabies scare in 1885, Pasteur concocted a treatment and gave the untested drug to a nine-year-old boy, Joseph Meister, who had been bitten by a grocer's dog. Three weeks later, Meister had almost fully recovered. Pasteur's legend received considerable help by the fact that Meister hailed from Alsace, a region controlled by Germany but claimed by France. The tricolor declared a victory for French science and for Pasteur who had beaten the German, Robert Koch, who had like Pasteur been working on vaccines. As a grown man, Joseph Meister took a job as a guard at the Institut Pasteur after Pasteur's death. When German troops entered Paris in 1940, they swarmed the institute's grounds and ordered that Pasteur's crypt be opened. Meister likely had been one of several men who defended the crypt against the Wehrmacht and prevented its defilement. Shortly after, Meister inexplicably shot himself through the head. Even this act became part of Pasteur's celebrity. Historians would write that Meister committed suicide in front of the Germans rather than disturb Louis Pasteur, France's hero.

Pasteur's influence on microbiology cannot be captured in a few pages. Early in his career, Pasteur had disproved the long-held theory of spontaneous generation, the belief that microbes and all other life arose from inanimate things: rocks, water, or soil. Biologists had already begun taking sides on this issue prior to Pasteur. As their science matured, many microbiologists doubted the logic behind spontaneous generation—science was increasingly distancing itself from spiritual dogma. Pasteur developed an experiment that unequivocally showed that a flask of sterilized broth could not produce life on its own. Pasteur modified this flask with an S-shaped tube to serve as the opening. This configuration let in air but prevented any particles from the air to enter. A second sterile flask left open to the air was soon teeming with bacteria but the S-flask remained sterile. Elegant in its simplicity, Pasteur's experiment earned him respect from his contemporaries.

During his career, Pasteur also distinguished between anaerobic and aerobic metabolism, invented the preservative method to be known as pasteurization, and developed the first rabies and anthrax vaccines (see Figure 2.3). As a postscript, the original S-flask is on display at the Institut Pasteur today and remains sterile.

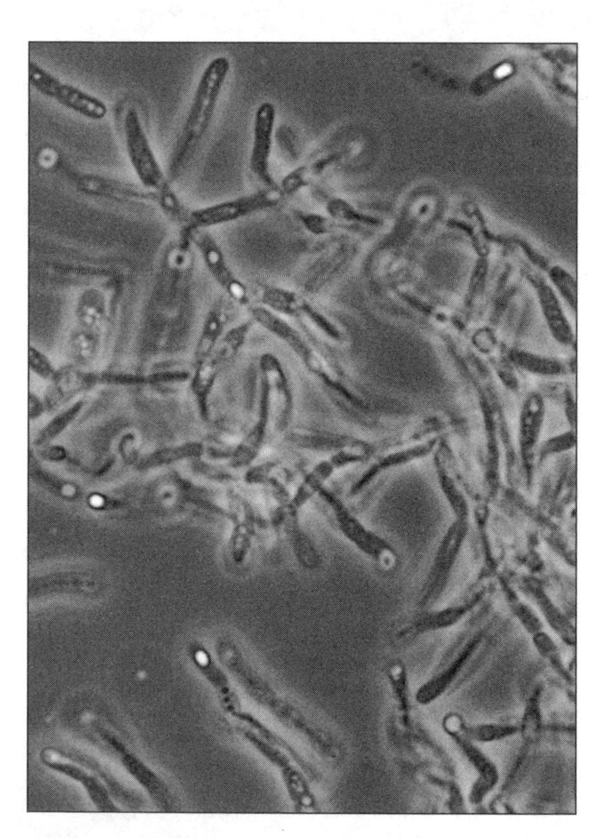

Figure 2.3 *Bacillus anthracis*, the anthrax pathogen. *B. anthracis* and all other *Bacillus* species form a tough, protective endospore. In this picture, endospores in phase-contrast microscopy look like bright ovoid balls inside an elongated cell. (Courtesy of Larry Stauffer, Oregon State Public Health Laboratory)

When bubonic plague erupted in Asia in the late 1800s, Pasteur dispatched Alexandre Yersin of the French Colonial Health Service to investigate. Microbiologists had by that time a century of using ever-improved microscopes, and they had become skilled at diagnosing disease by examining patient specimens to detect pathogens. In 1894, Yersin and a bacteriologist sent by Japan's government, Shibasaburo Kitasato, rushed with other public health officials to Hong Kong where a localized plague outbreak was emerging. Within a week Yersin isolated a rod-shaped bacterium from a plague victim. Kitasato found a similar microbe, but because the two men conversed only in broken German, they shared little of their findings. Yersin sent his report to the Institut Pasteur in Paris. Kitasato forwarded his results to Robert Koch in Berlin. In most circumstances, two scientists having attained the prominence in their profession as Pasteur and Koch had would have shared their data and drawn mutually agreed-upon conclusions. But Yersin and Kitasato's place in history would hinge on a rivalry between Pasteur and Koch that began 12 years earlier.

Pasteur and Koch held different perspectives on bacteria. Pasteur focused on the interplay between the body's immune system and bacterial pathogens and felt that virulence in pathogens changed over time, creating more or less virulent strains depending on environmental influences. Koch believed pathogens to be less variable and always capable of releasing virulence factors if opportunities for infection arose. Their differences would have made for lively and good-natured discussions had not Pasteur accidentally insulted Koch's "German arrogance" at a meeting in Geneva in 1882. Pasteur had actually praised Koch's body of work on anthrax and tuberculosis bacteria to the audience, but a scientist sitting next to Koch struggled to keep up with Pasteur's speech and translate it into German for his colleague. Unbeknown to Pasteur or Koch, the translator had made a mistake in going from French to German. In an age lacking telecommunications gadgets, the misunderstanding persisted. Koch returned to Berlin with contempt toward Pasteur that he made no effort to conceal.

When Pasteur published details of his successful rabies vaccine in 1885, Koch dismissed the work, insisting that a vaccine made of attenuated viruses needlessly endangered patients. But an underlying animosity likely arose out of each man's patriotism and the border conflicts between France and Germany over the Alsace and Lorraine regions. Koch undoubtedly remembered that Pasteur had received an honorary degree from the University of Bonn in 1868, but returned it later during the height of French-German tension. "Today this parchment is hateful to me," Pasteur wrote to the university dean, "and it offends me to see my name, which you have decorated with the qualification *virum clarissimum*, placed under the auspices of a name that will henceforth be loathed by my country, that of *Guillermus Rex*." The Germans responded with equal vitriol with both letters ending up printed in local newspapers.

With this history as a backdrop, Yersin and Kitasato hardly stood a chance of reaching an agreement on who had discovered the plague pathogen. Kitasato would unsuccessfully argue for the rest of his career that his discovery was the same as Yersin's, but Yersin received the accolades. He named the microbe *Pasteurella pestis* in honor of his boss. (Microbiologists still treasure the compliment of having a lethal pathogen named after them. The species would be renamed *Yersinia pestis* in 1944.) Historians have had trouble finding evidence in Kitasato's notes to confirm his discovery of the plague bacterium. A new generation of microbiologists would try to smooth the prickles by conceding that Kitasato had probably seen the same bacteria in his microscope that Yersin had spotted.

Pasteur and Koch never resolved their differences and Pasteur remained a French patriot to the end. In 1895, the Berlin Academy of Sciences extended a peace offering to Pasteur by inviting him to accept the medal of the Prussian Order of Extreme Merit. The Frenchman refused any invitation to Germany as long as it still held Alsace and Lorraine.

## Unheralded heroes of bacteriology

The names Pasteur, Lister, and Fleming represent significant advances in bacteriology but, as in today's technical fields, the rise to prominence results as much from personality as it owes to scientific merit. Generations of scientists since van Leeuwenhoek's day pursued the secrets of bacteria with the same devotion as more famous microbiologists. Many of their stories have been all but lost due to oversight, misunderstanding of their discoveries, and sometimes jealousy.

### *Robert Hooke*

In the 17th century, Robert Hooke corresponded with Antoni van Leeuwenhoek on the assembly of lenses for viewing the natural world on a microscopic scale. Both men developed similar instruments, but van Leeuwenhoek would become known as the Father of Microbiology while Robert Hooke's name has faded into near obscurity. A brilliant biologist and engineer, Hooke also mastered physics, the arts, architecture, geology, and paleontology over his long career.

As a youngster, a case of smallpox had disfigured Hooke, but he compensated with a gregarious nature. By the time Hooke graduated from Oxford, scientists in England sensed a luminary had arrived to raise public opinion of their profession. In 1662, the Royal Society of London elected Hooke at age 27 as Curator of Experiments, a role suited to his intellect and penchant for innovation. As Curator, Hooke performed an impressive array of demonstrations in biology, chemistry, and physics for the Royal Society but had an increasingly hard time staying focused on details from month to month. He often bolted to new projects before finishing the last, leaving other Society members to the drudgery of completing his studies.

Hooke tinkered with van Leeuwenhoek's microscope design and began a detailed study of the world he found under its lens. Hooke drew sketches of insects, feathers, plants, and leaves as well as snowflakes and mineral crystals and published them in *Micrographia* in 1665. (The book's full title is *Micrographia: Or Some Physiological Descriptions of Minute Bodies Made by Magnifying Glasses with Observations and Inquiries Thereupon.*) In it he coined the term "cell" to describe similar but separate units that composed a thin slice of cork. Overlooked at the time, this remark laid the foundation for all of biology: the cell is the simplest basic unit of every living thing on Earth, and without cells life does not exist.

The breadth of Hooke's accomplishments in architecture and engineering are no less impressive, yet Royal Society records contain little mention of the man or his work. In 1672, mathematician Isaac Newton out of Cambridge joined the Royal Society. Hooke had already begun developing equations to describe the gravitational forces of Earth's elliptical orbit around the Sun when Newton arrived at the Society with expertise in this same subject. The frail, introverted Newton developed a rapport with the outgoing Hooke as they pondered the mathematics of planetary movement. Their alliance permanently dissolved when in 1672 Hooke publically criticized a presentation to the Society by Newton on the properties of light and color. The animosity grew over the next decade. Hooke accused Newton of claiming credit for theories Hooke felt he had already developed. When in 1687, Newton published a thesis on planetary orbits with no mention of Hooke, the rift seemed irreparable.

Hooke's personality disintegrated in later years for reasons unknown. Isaac Newton would be one on a long list of people to whom Hooke directed his animosity. Newton took the Curator position shortly after Hooke's death in 1703 and almost at once struck Hooke's name from Society documents. Hooke's portrait disappeared under suspicious circumstances as well as many of his laboratory notes. Some historians believe those missing notes contain evidence that Hooke invented the compound microscope rather than van Leeuwenhoek, and questions persist on whether Hooke had developed the theory of gravity before Newton. Hooke would sadly become known as much for his rivalry with Isaac Newton as for his contributions to science.

### *John Snow*

Epidemiology owes its beginning to a London doctor's dogged attempt to stem one of several cholera outbreaks that had tormented London in the 1800s. Physician John Snow wrote in his journal in September 1854, "The most terrible outbreak of cholera which ever occurred in this kingdom is probably that which took place in Broad Street, Golden Square, and the adjoining streets, a few weeks ago." Snow's nonplussed colleagues knew of his tedious house-by-house assessment of family health and daily habits near the outbreak's center in Soho. The details he collected from the interviews seemed to have nothing to do, however, with the debilitating diarrhea that claimed many of the afflicted.

Snow persevered and sifted through his stacks of notes. He found that 73 of the outbreak's 83 deaths occurred within two blocks of a pump (see Figure 2.4) that dispensed water free to the public. The incidence of diarrhea related to the frequency in which families used the pump. By simply removing the pump's handle to make it unusable, Snow stopped the 1854 Soho cholera outbreak. He would become known as the Father of Epidemiology. Today's epidemiology follows the same path used by Snow. Epidemiologists track the locations where disease incidence are highest and search for commonalities among the sick. They gather other clues, such as an increased reporting by doctors and hospitals of common symptoms. Epidemiologists have even identified the presence of a waterborne outbreak by the uptick in sales of toilet paper in a community.

Snow conducted his study without any idea of the pathogen coming from the pump. His contemporaries had not connected water with many of the diseases of the day. Thirty years after the Soho outbreak, German microbiologist Robert Koch identified *C. vibrio* as the cause of the waterborne disease.

### George Soper

In 1883 Irish immigrant Mary Mallon arrived in New York City and found work cooking for well-to-do families. In the summer of 1906, Mary escaped the city heat and took a job at the rented cottage of banker Charles Warren in Oyster Bay, Long Island. Soon afterward Warren's family and staff suffered headaches, lethargy, loose bowels, and debilitating fever. The family doctor recognized the symptoms of typhoid fever but doubted an inner-city disease would afflict suburbia's wealthy.

Figure 2.4 The Broad Street pump. London has preserved the Broad Street pump as a historic site where John Snow stopped a deadly cholera outbreak. Prior to Snow's accomplishment, most doctors did not believe water carried disease. (Courtesy of Peter Vinten-Johansen, et al., *Cholera, Chloroform, and the Science of Medicine: A Life of John Snow*, 2003, 289; and http://johnsnow.matrix.msu.edu/images/online_companion/chapter_images/fig11-2.jpg)

By summer's end the Warrens had recuperated and returned to the city. The house's owner, George Thompson, heard of the outbreak and made a brilliant assumption: He suspected that a dangerous germ had entered his home. The Thompsons called on a public health officer, the fastidious, systematic, and humorless George Soper. Soper went straight to hands and knees at the Thompsons' in search of dirt, an undertaking depicted in the satirical cartoon in Figure 2.5. He sat for hours perusing household records and the comings and goings of staff and visitors. Soper scoured details that few epidemiologists had in the past. In a meal log, Soper noticed the Warrens' fondness for ice cream and sliced fresh fruit, excellent carriers of germs. He also noticed Mary's name in the records at the time the Warrens got sick. Soper hurried back to New York and unearthed health records showing that in seven of the eight families for whom Mary cooked, typhoid fever broke out; 28 cases in all and three deaths.

Figure 2.5 Health inspectors react to newspaper headline, "Looseness of the Bowels is Beginning of Cholera." (Courtesy of Wellcome Library, London; The John Snow Archive and Research Companion, Center for the Humane Arts, Letters, and Social Sciences online at Michigan State University)

Soper tracked down Mary the next year working in a Park Avenue apartment. With little formality he accused her of spreading death and disease and ordered her to surrender a fecal, urine, and blood sample on the spot. Husky and with a lightning temper, the cook hustled Soper out the door and into the street. Undaunted, he showed the city's Health Department his evidence and demanded action against the cook. Authorities felt Park Avenue was as unlikely a place for typhoid fever as Long Island, but Soper's meticulous notes swayed them. Police wrestled Mary out of the apartment and took her to Willard Parker Hospital, the main center for treating contagious diseases. There, doctors found unusually high concentrations of *Salmonella typhi* in her stool and the legend of "Typhoid Mary" was born.

Soper had all but called Mary Mallon an evil genius and put equal blame on upper class women who brought people like Mary into their homes. He irrationally likened them all to murderers. In 1928, he told the *New York World*, "She knew that when she cooked she killed people, and yet she deliberately sought employment as a cook." In fact, Mary Mallon never believed she had made anyone sick. Soper fought the prevailing beliefs of the day to stop the Typhoid Mary outbreaks. He also spearheaded the inspection of New York's sewers, water supply, and garbage pickup and became an advocate for good personal hygiene and community sanitation as the best ways to break the transmission of pathogens.

I thought of Mary Mallon in 1998 at a San Francisco crafts festival. Waiting in the lunch line, I noticed a young woman behind the salad-mixing station. During a lull in the action, she fished around in her mouth with her fingers and cleaned her teeth. She then plunged her unwashed, gloveless hands into a trough of lettuce, mixing the leaves and scooping out salad. I stepped out of line and said to her, "Do you realize you just contaminated all that salad with your dirty hands?" She looked confused at first then glanced at the lettuce. "Good," I thought, "I've taught someone about hygiene and averted a possible health disaster."

In 1909, New York quarantined Mallon on an island in the East River. Miserable, angry, and convinced she had nothing to do with typhoid, she lamented her role as "a peep show for everybody." After her release Mary changed her name and began cooking again, this time at Sloane Maternity Hospital. By 1915, she had caused 25 new cases of typhoid until a health inspector spotted her in the hospital kitchen. Police took Mary back to the island where she died in 1938.

Unlike other pathogens, *S. typhi* lacks multiple strains of varying virulence; the species is fairly uniform the world over. *S. typhi*'s survival rests on asymptomatic carriers who efficiently spread the pathogen through a population. Research has not yet uncovered all of the secrets of typhoid susceptibility in those who are asymptomatic. In carriers the bacteria multiply in the gallbladder, bile duct, and intestines, and then spread in drinking water and food contaminated with miniscule bits of fecal matter, a more common occurrence than most people think. I had no more success in convincing the salad vendor to change her hygiene habits than Soper had with Mallon because no one wants to believe they disseminate contamination.

Because of the prevalence of fecal bacteria everywhere, people would be wise to take a bacteriocentric view of the world, "seeing" bacteria on the places they exist even though they remain invisible. Suspicious foods, dirty floors, or murky water shout the presence of bacteria as do people who avoid washing their hands. In the 1970s, bacteriocentricity helped solve one of modern epidemiology's most puzzling outbreaks.

### *Joseph McDade*

Mid-July in Philadelphia is sticky, sweaty, and heavy with odors that seem to seep from the concrete. In 1976, the 70-year-old Bellevue Stratford on South Broad Street opened its doors to 4,000 World War II Legionnaires in town for their annual convention. An influenza outbreak that had killed a soldier in nearby Fort Dix, New Jersey, that summer put many of the visitors on edge, especially because the virus resembled one that took 40 million lives in 1918 to 1919, the worst single flu outbreak in history. The Legionnaires and hotel staff likely took extra care washing their hands and staying on guard for the sounds of sneezing or coughing. But trouble came from a different direction.

The Bellevue Stratford's air conditioning system had developed a thick biofilm in the condensation-wetted distribution lines. There, amoeba, a type of protozoa, multiplied in the moist habitat they need for survival, feeding on the biofilm. Hidden inside the amoeba lived a bacterium that most microbiologists did not know existed. Because of the biofilm, the air conditioning vents began to emit moisture droplets filled with microbes.

No one knew they were inhaling contaminated air. Hotel guests and even people who had strolled past the building's open doors became congested and weak, developed muscle pain and headaches, and suffered with diarrhea. The dreaded flu virus obviously had returned and with it, near panic—someone blamed the communists. Congress ordered into place an emergency vaccination program, but the year closed with few weapons against the mysterious disease.

During the holidays, Joseph McDade from the Centers for Disease Control and Prevention (CDC) stared into his microscope searching blood samples from the hotel's guests for *Rickettsia* bacteria. *Rickettsia* bacilli would be easy to miss because it lives only inside other cells, such as human cells. Weary with eyestrain he went to a holiday celebration, but for McDade, bacteria were more compelling than office parties. He returned to his laboratory and re-examined the Legionnaires' samples. In the early hours he spotted a cluster of bacilli inside white blood cells. The bacteria were not stubby, short rods of 1 µm like *Rickettsia*, however, but long thin rods stretching to 10 µm or more.

McDade had found a new species, *Legionella pneumophila*. The CDC unraveled the bacteria's pathology. *L. pneumophila* enters the lungs and then infects the bloodstream. The immune system releases cells called macrophages for the specific purpose of destroying infectious agents such as bacteria, but like *Rickettsia*, *Legionella* is a "stealth pathogen." *L. pneumophila* slips inside macrophages and multiplies in the phage's cytoplasm. A new generation of cells bursts free and continues the infection cycle. Microbiologists had noted bacteria that fit *L. pneumophila*'s description years earlier, but the microbe's finicky growth requirements made laboratory studies almost impossible.

Clinical microbiologists deal with a short list of stealth pathogens in addition to *Rickettsia* and *Legionella*, including the foodborne pathogens *Listeria monocytogenes*, *Shigella flexeri*, and *Salmonella enterica*, and mycoplasmas. *L. monocytogenes* invades the epithelial cells lining the digestive tract, and when in the bloodstream is one of the few bacteria that cross the blood-brain barrier. Severe cases of listeriosis therefore damage the central nervous system. *Salmonella* and *Shigella* usually stay in the digestive tract.

## On the front

Microbes have played a part in war before the Tartars deployed their unique bioweapon. Prior to the introduction of antibiotics, minor battlefield injuries led to about half of all wartime deaths. Marginal food, lack of sleep, and emotional stress reduced soldiers' ability to fight infection. Without treatment of infected wounds, pathogens could enter the bloodstream and multiply—a condition called sepsis—and then infect major organs. Some pathogens stay at the wound site and cause severe infection there. Badly injured skin contains oxygen-free pockets in the tissue, which promotes the growth of anaerobes such as *Clostridium perfringens*, the cause of gas gangrene. Before World War II small scratches caked with soil and left untreated presented the risk of amputation or death.

Virulence factors aid the infection process. Some bacteria rely on only one approach, such as *Mycoplasma* that produces hydrogen peroxide and ammonia, both toxic to the body's cells. After the two compounds damage cells lining the respiratory tract, *Mycoplasma* enters lung tissue. *Staphylococcus aureus*, by contrast, uses a battery of weapons:

- Coagulase enzyme clots the blood surrounding a wound and protects the bacteria from the body's immune defenses.
- Nuclease enzyme breaks up exudates in the wound and thus helps the bacteria's mobility.
- Hemolysins lyse red blood cells, causing anemia and weakened body defenses.
- Hyaluronidase enzyme degrades the binding material between human cells to aid passage of the pathogen throughout the body.
- Protein A binds the body's antibodies and renders them inactive.
- Streptokinase enzyme activates a series of steps in blood clot destruction, allowing the bacteria to escape a clotted area.

Two champions of proper medical care died a few years before the First World War. British nurse Florence Nightingale called for reforms in treating combat injuries. During her service in the Crimean War, she reported on the diseases, poor food, and unsanitary conditions in medical hospitals. Her 1,000-page report compiled in 1858 convinced her superiors that the British Army was needlessly losing soldiers to treatable injuries. During the same period in Britain, surgeon Joseph Lister insisted that surgeries required sterile conditions and wounds must be kept clean with antiseptics. Lister used carbolic acid as an antiseptic; several years would pass before less irritating chemicals came into use.

Sterility and antiseptics were new ideas when war began in 1914. Not all surgeons wanted to put chemicals on patients' skin, and they initially resisted using antiseptics. A second, more revolutionary, defense against infection soon surfaced. Microbiologist Felix d'Herelle had tried to fight a locust outbreak by infecting the insects with bacteria. He presumed that if a similar agent attacked pathogenic bacteria, it would fight infectious disease. d'Herelle knew that some microbiologists had discovered a substance in their bacterial cultures that infected and killed other bacteria. With little idea of the material's identity, d'Herelle began collecting liquid medium from affected cultures. By 1917, he was using it to cure hundreds of cases of dysentery by injecting patients with his "antagonistic microbe." Not until 1939 with the advent of electron microscopes did microbiology learn of bacteriophages, viruses that infect only bacteria. The treatment called phage therapy would be superseded by antibiotics in the next decade, but for a short period in history phages played the part of the magic bullet.

The consequences of war, upheaval of home life, and the creation of mass refugee migrations hampered sanitation and personal hygiene. The common body lice *Pediculus humanus* infested almost everyone in World War I. The lice carried the typhus bacterium *Rickettsia prowazekii*. This microbe behaves like a virus by living as a parasite inside other cells. The lice ingested *R. prowazekii* when they bit an infected person and after an incubation of six days they became infective to others. Unlike the plague, which spread via flea bites, typhus spread when lice defecated on the skin and the bacteria entered the body through a wound.

Typhus would blanket Europe and become an epidemic second only to the Black Death in fatalities. In Serbia, 20 percent of the population contracted typhus and 60 to 70 percent of those people died. Disease became so devastating to Austria, the Balkans, Russia, and Greece that the Central Powers delayed some maneuvers for fear of wiping out their own armies. At the close of the war, a four-year epidemic struck Russia and would kill half of the 20 million people who had been infected with typhus.

Many in the German army that invaded Poland to start World War II carried memories of the typhus outbreaks. Entering the third year of occupation, Polish physicians Eugene Lazowski and Stanislaw Matulewicz devised a way to stop some of the carnage and deportations to work camps. They knew that the *Proteus* strain OX19 looked similar to *R. prowazekii* to the body's immune system. They thus began injecting healthy residents of the town of Rozvadow with killed OX19 cells. This ersatz vaccine induced the production of antibodies against the typhus bacterium. Lazowski and Matulewicz had created a fake typhus epidemic.

The Germans may have had suspicions of the isolated outbreak. A German medical team arrived in Rozvadow in 1942 to assess the situation, but their doctors so feared infection that they skipped giving physical exams; they collected blood samples and hurried back to Berlin. The antibodies in the samples convinced the German army to avoid typhus-ridden Rozvadow. The contrived typhus epidemic saved almost 8,000 lives, many of them Jews.

People and their pathogens have continuously traded victories and defeats. Sometimes the bacteria win, such as in plague and syphilis epidemics. Sometimes the guile of humans triumphs as in d'Herelle's phage therapy. But do people ever truly defeat bacteria? The search for the magic bullet ended when a shy microbiologist discovered the "miracle drug" penicillin, or so it seemed at the time.

# 3

## "Humans defeat germs!" (but not for long)

In bacteria, one mutated cell appears for every 100 million normal cells. Because some bacteria reproduce as quickly as every 20 minutes, new populations of mutants emerge literally overnight. Most mutations give no discernible advantages or disadvantages to the cell. Unfavorable mutations make bacteria vulnerable to other microbes or the environment, and these cells and their genes disappear forever. On rare occasions a mutation gives a bacterial cell a favorable characteristic called a trait that enables the bacterium to outperform others.

Most people remember from Biology class that a favorable mutation appears only because of a random event. "The survival of the fittest" comes not by plan but by luck. Chance mutations in bacterial DNA produce slight, random changes in a single gene, and this altered gene gives the cell the ability to grow faster, swim farther, absorb more nutrients, or withstand heat better than its brethren. When this special cell divides, two identical cells appear that also outcompete others until the new gene has become part of a new, evolved population.

In 1988, John Cairns found in *E. coli* a ploy that turned the concept of randomness on its head. Cairns's *E. coli* used adaptive mutations, which occurred when a specialized mutator gene detected a stimulus in the environment. Mutator genes prompt the cell's mutation rate to speed up, thus increasing the chance that one of *E. coli*'s 4,377 genes will mutate in a favorable direction. More than 30 mutator genes have now been located in *E. coli* and similar genes in *Pseudomonas aeruginosa*, a water-associated microbe and common invader of burns and invasive devices (intravenous tubes, catheters, and so on). Are bacteria choosing how and when they mutate? If so, an idea that once belonged only in science fiction may be a reality.

## What is an antibiotic?

Antibiotic means "against life" and belongs to two groups: true antibiotics and bacteriocins. A true antibiotic is made by a microbe to kill other unrelated microbes. *Penicillium* mold produces the antibiotic penicillin to kill bacteria that venture too close to its territory. Bacteriocins come from bacteria to kill other bacteria. For example, *E. coli* produces the bacteriocin colicin that kills bacteria in *E. coli*'s family of enteric microbes. Some bacteriocins kill different strains of the very same species, all for the purpose of reducing competition for space, food, light, and water.

An antibiotic that kills bacteria outright is a "cidal" agent, or bactericidal. Weaker antibiotics that merely slow down bacterial growth are called bacteriostatic. Penicillin is bactericidal because it prevents susceptible bacteria from building a rigid cell wall, forcing the bacteria to succumb to toxins in their environment. Tetracyclines, by contrast, interfere with protein synthesis, which may not necessarily kill the cell. The cells might switch to an alternate synthesis pathway, but this slows their reproductive rate, so tetracycline has done its job. Figure 3.1 illustrates a simple laboratory test that determines the susceptibility of bacteria to various antibiotics.

The structure of an antibiotic includes several carbons and hydrogens plus carbon rings and branches that make the molecule look complex. Nature developed the intricate structures to make it harder for bacterial enzymes to recognize and degrade an antibiotic. But humans interfered with nature's plan by using increasing amounts of antibiotics and thus exposing bacteria to the compounds more frequently. Twenty years after the first commercial use of penicillin, antibiotic-resistant bacteria emerged. Resistant bacteria now exist for all of the natural antibiotics in Table 3.1. Today chemists try to stay ahead of bacteria by synthesizing new antibiotic molecules with more complexities in the hope of outwitting pathogenic bacteria.

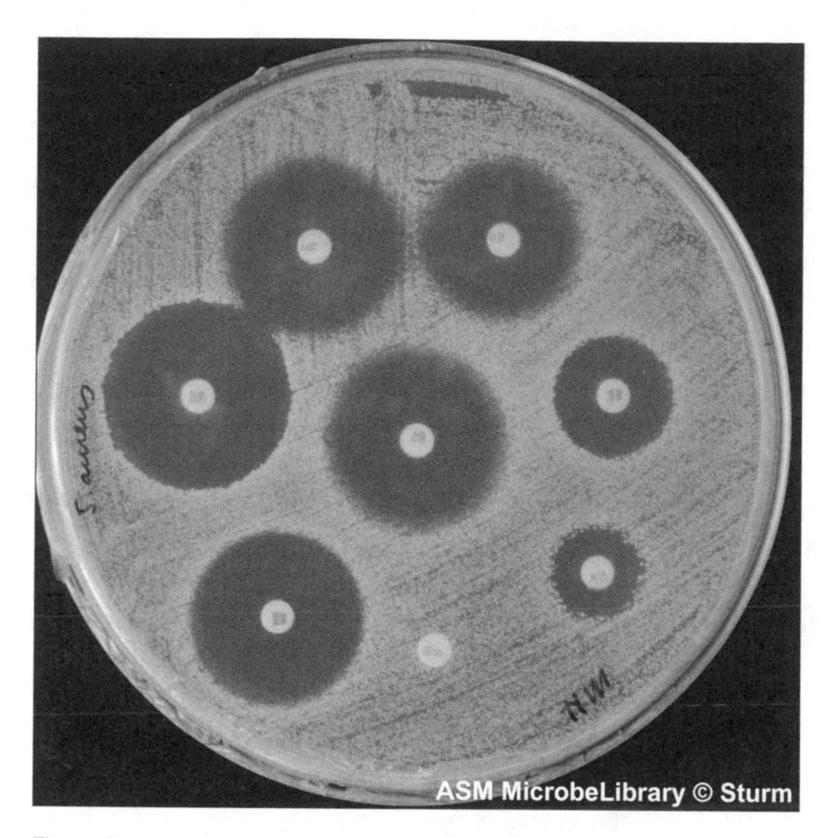


Figure 3.1 Kirby-Bauer antibiotic testing. Small paper discs soaked in different antibiotics cause varying levels of inhibition against bacteria. This test is a refinement of Alexander Fleming's discovery that mold spores can kill bacteria due to the secretion of antibiotic. (Reproduced with permission of the American Society for Microbiology MicrobeLibrary, www.microbelibrary.org)

**Table 3.1** The main natural antibacterial antibiotics

| Producer | Antibiotic |
|---|---|
| Molds | |
| *Acremonium* | Cephalothin |
| *Penicillium* | Griseofulvin and penicillin |
| Bacteria | |
| *Bacillus* | Bacitracin and polymyxin |
| *Micromonospora* | Gentamicin |
| *Streptomyces* | Amphotericin B, chloramphenicol, erythromycin, neomycin, streptomycin, and tetracyclines |

The United States produces 25,000 tons of antibiotics annually. Most of the drugs go to human medicine and agriculture. Cattle, hogs, sheep, goats, and poultry raised for meat receive 70 percent of the supply to promote growth rate and repel infections that travel fast through factory farms. The remainder of the antibiotic supply goes to dogs, cats, horses, and other domesticated animals, pelt animals, fish, and plants and trees.

Meat producers have suffered strident criticism for giving the animals they raise a constant intake of antibiotics. When I began my college career as an animal science major, we took for granted the benefits of antibiotic use in meat animals. Beef, pork, and poultry received subtherapeutic levels of more than one drug for no specific reason other than the possibility of increased weight gain. The mounting questions regarding this practice spurred researchers to study bacteria in the digestive tract of healthy ruminant and nonruminant animals receiving antibiotics. Antibiotic-resistant bacteria have been recovered from these animals, but it can be difficult to prove that the antibiotics led to resistance.

Food producers insisted for years that antibiotics are needed for efficiency in meat production. Meat producers give these drugs to animals to prevent the spread of infection in a population of animals living in very close quarters from birth to slaughterhouse—this is the reason behind the term "factory farming." Factory farming increases the nebulous condition we call stress when animals spend their entire lives squeezed together, and stress weakens immunity. The high density of individuals creates a higher risk for infection. Perhaps the logic behind factory farming and administering antibiotics to livestock makes no sense, and halting both would be a better choice.

The agriculture industry argues that efficient mass-production style farming keeps food costs low. Researchers have discovered shifts in the proportions of intestinal bacteria in antibiotic-fed animals. It has been more difficult, however, to determine the connection between altered bacterial populations and faster growth in animals.

Large-scale agriculture has been reluctant to share its antibiotic methods, so the public will have a hard time learning which antibiotics, if any, are in the meat they buy.

The environmental effects of subtherapeutic antibiotics in meat animals remain largely unknown. Two outcomes seem likely, however. First, antibiotic-resistant bacteria shed in manure enter the environment and cause harmful consequences in ecosystems, and second, eating rare meats or runny eggs increases a person's chance to ingest resistant bacteria. Food is not sterile and cooking does not guarantee the removal of all potential pathogens; cooking reduces bacterial numbers to safer levels. We get away with ingesting a pathogen here or there throughout the week because the dose of the microbe is lower than required to cause infection. At the same time, our native bacteria and immune system protect the body from exposure to low numbers of pathogens.

The European Union and Canada ban antibiotic use in meat animals, and the World Health Organization has taken a stance that conveys its concern over antibiotics in agriculture. The United States still uses antibiotics, and meat-producing states continue to contend that no indisputable evidence exists to prove that meat antibiotics lead to drug resistance in people. Indisputable evident is exceedingly difficult to find in any field of science, so consumers have been left with making their own decisions on the safety of meat products.

Antibiotics that escape farms in runoff from manure piles enter surface waters. In a perfect world, the wastewater would be directed to a wastewater treatment plant without contaminating the environment. This is impractical considering the magnitude of daily manure output in the United States alone. Wastewater treatment and drinking water disinfection provide poor protection against antibiotics. In 2005, researchers from the University of Wisconsin detected six antibiotics in treated wastewater:

- **Tetracycline**—For skin, urinary tract, and some sexually transmitted diseases
- **Trimethoprim**—Childhood ear infections, urinary tract infections
- **Sulfamethoxazole**—Used in combination with trimethoprim for treating ear, bronchial, or urinary tract infections
- **Erythromycin**—Treats respiratory tract infections

- **Ciprofloxacin**—Lower respiratory, urinary, and other infections
- **Sulfamethazin**——For respiratory and other infections in animals

I am not suggesting that treated water is a source of danger in every community or that antibiotics in water definitely cause harm. The drugs in the study described here had, furthermore, been detected in parts per billion levels, equivalent to one corn kernel in a nine-foot silo of corn.

Drugs entering the environment year after year are affecting ecosystems, but scientists do not yet know all the details. Therefore, the public has no way of knowing. But imagine an antibiotic injected into a sick horse ending up miles away in a glass of tap water or a plate of oysters Rockefeller.

Antibiotics made immediate and profound effects on human health when they first became available, and few people foresaw trouble. Trouble would come, and the first warning came from a surprising source. Unfortunately, the world missed the message concerning antibiotic resistance until it was too late.

## Inventing drugs is like making sausage

In 1897, a 23-year-old doctoral student submitted his graduation thesis to the Institut Pasteur, alluding to a new drug that might be helpful in fighting bacterial infections. The document described a *Penicillium* mold that killed *E. coli* in Petri dishes and cured laboratory animals injected with live typhoid bacteria. The reviewing faculty found Ernest Duchesne's work uninspired, but they granted Duchesne a diploma along with little encouragement for a career in science. He enlisted in the French army. Before leaving, Duchesne discarded his laboratory notes; his thesis disappeared into a corner of the institute.

World War I mimicked all prior wars by costing millions of lives from infections, many of them minor, received on the battlefield. On the front, nurses stretched their bleach supply by diluting it until it had no effect against any germ. About half of the war's 10 million fatalities came from infections. Duchesne did not get the chance to alert the world of his anti-infection drug. He caught tuberculosis soon after joining the French army and died at age 37 in 1912.

Another medical student in Germany had already begun his own hunt for a "magic bullet," a drug to kill pathogens without harming the patient. Paul Ehrlich tested 605 different substances in an effort to find a drug that killed many different types of pathogens but did not cause harmful side effects in patients. When he tested the arsenic-containing compound, salvarsan, he found it inhibited the *Treponema* bacteria that cause syphilis. The promising new drug became known as Compound 606. Prior to the discovery of salvarsan as an antibiotic, Western medicine depended on an antibacterial substance that Spanish conquistadors had learned of in South America. Peru's Quechua Indians had been using an extract from the cinchona tree to treat "ague." In the mid-17th century, Jesuit priests brought the Peruvian powder to Europe. The substance to become known as quinine caused little stir in the medical community until it cured England's Charles II of ague, now known as malaria. The new drugs energized physicians, biologists, and chemists toward finding other disease-curing compounds hidden in nature.

Chemists soon emulated Ehrlich, whom they had nicknamed Doctor 606, by testing hundreds of synthetic compounds against bacteria. In the early 1900s, however, chemical companies had little practice in drug research. Their chemical stockpiles were limited to fabric dyes for protecting threads against decomposition by bacteria. The compounds did not work well in laboratory tests against bacterial cultures, and in later years most of these substances were shown to cause cancers. Ehrlich would not realize his dream of finding a single magic bullet to kill all infectious disease.

Sixteen years after Duchesne's death, Scottish microbiologist Alexander "Alec" Fleming prepared for a short September vacation from his lab at London's St. Mary's Hospital. Historians have shaped the ensuing tale. Fleming had a reputation as a dedicated scientist but terrible housekeeper. His lab overflowed with Petri plates, tubes, beakers—certainly the makings of a contaminated experiment. While Fleming was away, rogue mold spores contaminated Petri dishes filled with *Staphylococcus* bacteria. When Fleming returned, he noticed clear zones in the film of *Staph* cells where a spore had landed, and he concluded that the mold had disintegrated the bacteria. No one is sure where the mold originated. Spores likely drifted from a floor below where mycologist C. J. La Touche's laboratory was chock-full of molds. Fleming's habit of messiness gave the spores plenty of places to land and grow.

More than one stroke of luck converged to propel Alexander Fleming into history. The early-fall temperatures were warm enough for bacterial growth but cool enough for mold contaminants such as *Penicillium*; *Staph* cells prefer body temperature while molds prefer room temperature. Fleming had been studying *Staph* cultures, which are particularly susceptible to the action of *Penicillium*. Perhaps the most fortuitous break occurred when lab assistant D. Merlin Pryce came by for a casual hello and spotted the *Penicillium*-inhibited *Staph* cells among the cultures.

To Alec Fleming's credit he investigated odd occurrences that others might dismiss as aberration. He continued studying *Penicillium*. Fleming had assumed that the mold had lysed the bacteria when spores landed on the bacterial film. Only later did microbiologists learn that *Penicillium* targets young, growing bacteria. The mold spores had probably contaminated Fleming's *Staph* cultures before he began his vacation.

Fleming published his results in 1929 and gave lectures on the new substance he called penicillin. But because of acute shyness, Fleming reduced the most riveting topics to a monotonous drone, and he failed to inspire his peers. His colleague at St. Mary's, pathologist Almroth Wright, openly disparaged Fleming's work. Alec Fleming retreated to his lab and his main interest, a new compound called lysozyme that he had discovered in human tears. (Fleming developed most of the knowledge biologists have today on lysozyme. This enzyme serves as a first-line defense against pathogens on the skin or near the eyes. Fleming's important discovery would be overshadowed by his research on penicillin.)

When the British entered World War II, German bacteriologist Gerhard Domagk had already discovered sulfa drugs. Britain's doctors saw the advantage these drugs gave the German infantry for treating wounds, but their own laboratories offered nothing similar. In 1938, Oxford University pathologist Howard Florey had teamed with a recent refugee from Germany, Ernst Chain, to find an anti-infection drug for Britain. Chain unearthed Fleming's 1929 article on the effect of mold on *Staph*, and the two suspected they had a diamond in the rough. Florey and Chain extracted penicillin from the mold, and then began the lengthy, tedious task of purifying and scaling it up to useful amounts. Back in London, Fleming alternated penicillin experiments with lysozyme studies. During the London Blitz he expanded the list of bacteria susceptible to penicillin and designed clever tests to differentiate mildly susceptible bacteria from the highly susceptible.

In late 1940, Florey and Chain published a brief article in a medical journal on a *Penicillium* extract hundreds of times stronger than Domagk's sulfa drugs in killing gas gangrene *Clostridium*. Not until August of 1942 did the *London Times* pick up the story, but it mentioned no scientists by name. Almroth Wright who had so harshly criticized Fleming 13 years earlier pounced on an opportunity. He wrote the *Times* to inform them of penicillin's discoverer Alexander Fleming, with special credit to St. Mary's Hospital. The headlines "Professor's Great Cure Discovery," "Miracle from Mouldy Cheese," and "Scottish Professor's Discovery" began appearing. St. Mary's hospital basked in the recognition (and the increased donations) that other London hospitals coveted.

The public had never heard of Florey or Chain, but Fleming and the scientific community kept abreast of their attempts to scale-up penicillin production. In August, Fleming, who had never developed the knack for making large quantities of purified penicillin, asked Florey for some of his drug to treat his friend Harry Lambert, suffering with a severe streptococcus infection. Florey rushed to London with his entire stockpile of pure penicillin and showed Fleming how to inject it. Although Fleming bungled Florey's instructions, he nonetheless saved Lambert from certain death.

Everyone now wanted to know about the new drug, and Fleming may have felt obligated to offer uplifting news. He hinted at penicillin's promise for saving the lives of Britain's troops. Florey knew better. Britain had reached the limits of its manufacturing capacity. In his view, Fleming and St. Mary's Hospital reaped publicity and donations based on false hopes. Between air raids, Florey and colleague Norman Heatley had been scrounging jars, bottles, even bedpans to keep up with the demand for new batches of penicillin. In 1941 both men obtained coveted tickets for Pan Am's Dixie Clipper flight across the Atlantic. On the trip Florey carried a briefcase stuffed with mold cultures and a handful of vials of pure penicillin with the hope to get help from a large American drug company—they visited Merck, Pfizer, E. R. Squibb, and Lederle Laboratories—for mass-producing penicillin. As late as 1942, Britain's version of mass production involved collecting *Penicillium* extract in bathtubs, and then rigging milking equipment for the purification steps.

Florey's campaign for penicillin took a lucky turn when he visited Yale medical researcher John Fulton during his second year in the United States. Fulton told Florey of a local woman, Anne Miller, who had been dying with a seemingly incurable *Streptococcus* infection. Fulton had cajoled a few grams of penicillin from Merck in New Jersey, which Florey had visited the previous year. At 3:30 in the afternoon on a cold March day in 1942, Miller had been consigned to death with a fever over 100 degrees when she received her first dose of penicillin. By 4:00 the next morning her temperature had returned to normal. Miller's recovery shocked even Fulton. He preserved her hospital charts, which now belong to the Smithsonian Institution. By the close of the war, American drug companies were producing 30 pounds of penicillin a year, enough to treat a quarter-million patients for a month.

In his acceptance speech for the 1945 Nobel Prize in medicine Fleming shared with Florey and Chain, he commented on the future of antibiotic drugs. Perhaps, Fleming mused, a time would come when anyone with real or perceived illness could get penicillin. "The ignorant man," he warned, "may underdose himself and by exposing his microbes to non-lethal quantities of the drug, make them resistant." Fleming described a hypothetical scenario of resistant bacteria infiltrating families, and then entire communities. On that December day the story of penicillin's discovery captured the public's imagination more than the remote oddity of resistant bacteria.

## Mutant wars

Alec Fleming's fear of antibiotic overuse and misuse soon became reality. Doctors began prescribing antibiotics for minor injuries, headaches, colds, flu, and other ailments. Even perceptive physicians who worried over indiscriminate use of the drugs could be badgered into prescribing them by patients who felt lousy. The patients did not know or perhaps did not care that antibiotics had no effect on colds, the flu, and other viral infections.

In the 1960s, rather than slowing down to do more research on antibiotics, agriculture stepped up the use to fight imaginary infections and put more weight on livestock or plump poultry before sending them to market. Resistant bacteria began showing up in places in addition to hospitals. A microbiologist taking a sample of bacteria from a person's digestive tract, skin, or mouth, or from natural waters and soil had a very good chance of finding more than one resistant species. Antibiotic-resistant bacteria now settle on kitchen counters, gym equipment, and in locker rooms. Franz Reinthaler showed in 2003 that antibiotic-resistant *E. coli* exists at every step in wastewater treatment, and most of the strains tested have resistance to more than one antibiotic. The microbial world has become almost saturated in antibiotics and thus in antibiotic-resistant microbes.

Bacteria excel at adaptability. Bacteria carry genes that confer antibiotic resistance in their large DNA molecule, the chromosome, and also on small circular strands of DNA called plasmids that stay separate from the chromosome in the cytoplasm. Resistance genes give bacteria the ability to fight antibiotics in five ways: (1) by cleaving antibiotics into pieces, (2) blocking an antibiotic's penetration into the cell by altering the drug's normal entry site, (3) pumping the antibiotic out of the cell as soon as it penetrates, (4) repairing any damage the drug does inside the cell, or (5) altering metabolism to lessen the antibiotic's damaging effects. Put another way, bacteria have at least as many tactics for resisting antibiotics as antibiotics have modes of action.

Penicillin, sulfa drugs, and the other new antibiotics introduced in the 1940s and 1950s delivered some remarkable cures. Doctors treating very sick patients were probably tempted to try antibiotics on nonbacterial diseases in the hope that the drug helped kill secondary infections even while they knew the primary infection had been caused by a virus. If a doctor noticed that an antibiotic began losing strength, he would simply prescribe a new antibiotic. Sometimes a patient received both drugs at the same time. Two antibiotics together stymied bacteria for a while, but this strategy created problems. Any two random antibiotics cannot be paired and expected to work better than either drug alone. Certain antibiotics lower the activity of the second: streptomycin inhibits chloramphenicol's activity; erythromycin blocks penicillin's activity. When tetracycline acts on *Staphylococcus*, for instance, it inhibits protein synthesis in mature cells. But penicillin requires new, growing cells to exert its activity against the cell wall. By slowing bacteria's growth, tetracycline neutralizes penicillin's mode of action.

Multiple antibiotics, even when paired correctly, also led to multidrug resistance. Bacteria now evade many antibiotics at the same time. This is not an extraordinary talent as nature already exposes bacteria to more than one antibiotic or bacteriocin at a time, and multidrug resistance probably already existed in a minority of bacteria. Soil bacteria face a dense community of antibiotic-producing fungi and bacteria that make antibiotic resistance essential for survival. The proliferation of antibiotic use from the 1950s to the1980s merely accelerated the evolution of antibiotic defenses.

Some bacteria began carrying additional resistance genes for more than one antibiotic. Methicillin-resistant *Staphylococcus aureus* (MRSA) has separate genes that control resistance for the penicillin family of antibiotics as well as genes for resisting tetracycline, clindamycin, aminoglycoside, and erythromycin.

Bacteria with pump mechanisms could eject an antibiotic as soon as the drug passed through the cell wall and membrane. These bacteria developed more sophisticated systems to resist multidrug treatment with an adaptation called the ABC transporter, for "ATP-binding cassette transporter." (A cassette is a set of genes that work as a team.) Present in bacteria, archaea, and eukaryotes, ABC transporters are proteins that help pump certain harmful molecules out of the cell. (Cancers that do not respond to chemotherapy resist the treatment in part by employing ABC transporters to eject the drug from tumor cells.)

ABC transporters consist of two proteins that span the bacterial membrane from the inner surface surrounding the cytoplasm to the membrane's outer surface. The two proteins form a pore through the membrane. By expending energy, the cell uses this pore to expel a variety of chemicals, including more than one type of antibiotic. About 30 different types of ABC transporters exist among bacteria to eject from cells the diverse chemicals that can harm them in their environment. In addition to antibiotics and bacteriocins, transporters carry bile salts, immune system factors, hormones, and carried chemicals called ions, and they have recently been shown to adapt to, and eject, human-made antibiotics.

Multidrug resistance among bacteria has now become more prevalent than resistance to a single antibiotic. Some bacteria carry so many defenses it seems as if they were designed specifically to defeat drug companies' best efforts. The tuberculosis bacterium *Mycobacterium tuberculosis* contains 30 different ABC transporters that provide the species with a defense that acts as a backup to other defensive schemes. First, the microbe's unusual cell wall composition prevents the penetration of many antibiotics that work on other bacteria. The ABC transporter system acts on any antibiotic that manages to get past the cell wall. Second, *M. tuberculosis*'s capability to hide inside cells of the immune system enables it to elude antibiotics circulating the bloodstream. Third, these bacteria grow like the tortoise compared with *E. coli*'s hares. Slow growth may not in itself be a defensive tactic, but this characteristic of the species forces doctors to lengthen the antibiotic treatment for TB. Because most antibiotics work best on actively dividing bacteria, *M. tuberculosis*'s growth rate lessens an antibiotic's killing efficiency. Typical TB treatment lasts six months or longer, and this alone favors the pathogen because even diligent patients have a hard time staying on a drug regimen for that long.

The multiple defenses of *M. tuberculosis* necessitated more than one antibiotic when doctors began treating the disease with antibiotics in the 1940s. Two antibiotics worked well for many years, but now this species requires four different drugs to kill it, and many strains already resist all four, leaving doctors with a dwindling choice of antibiotics that still work against TB. Like other bacteria, when *M. tuberculosis* has developed a favorable trait, it keeps the gene for that trait in its DNA. Multidrug resistance has also become common in skin infections, sexually transmitted diseases, and pneumonia.

Following Germany's 1936 introduction of sulfa drugs to cure gonorrhea, resistant strains of *Neisseria gonorrhoeae* had spread throughout the country by 1942. Doctors turned to penicillin as soon as U.S. drug companies made large quantities available. Before the 1960s had arrived, resistant *N. gonorrhoeae* capable of cleaving penicillin into pieces had spread around the globe. Almost all *Staphylococcus* species had already become resistant to penicillin 15 years earlier. Bacteria have become so efficient in building and sharing resistance that they no longer need months or years to adapt. Four days after streptomycin therapy begins, for a kidney infection for instance, streptomycin-resistant bacteria outnumber the susceptible bacteria in patient urine samples.

Bacteria possess an effective defense against antibiotics: the plasmid. Bacteria of the same species or sometimes dissimilar species pass plasmids back and forth and thereby give each other useful traits they would not normally possess. Sometimes bacteria insert a resistance gene from their chromosome into a plasmid before passing the plasmid to other cells. Cells also share entire DNA segments from the chromosome by absorbing pieces when another cell dies and breaks apart or by connecting cell-to-cell in a version of bacterial sex.

Microbiologists have tried various approaches to outmaneuver bacterial defenses against antibiotics. One ploy called the Trojan Horse takes advantage of the competition for iron among living things in nature. Because iron can be scarce in many habitats, bacteria produce compounds called siderophores to seize hold of precious iron molecules and bring the metal into the cell through a specific pore. Microbiologists have designed siderophores that instead of grabbing iron will bind to an antibiotic. When bacteria recognize the siderophore, they open the pore to let it in and thus allow the antibiotic to enter.

If certain bacteria do not fall for the trick of smuggling an antibiotic into their cells, microbiologists try substituting the metal gallium for iron in siderophores—the two metals look similar to bacteria—to derive the bacteria of essential iron.

Microbiologists have another weapon at their disposal in bacteriophages or phages, which are viruses that attack only bacteria.

In the microbial world, bacteria look like the mother ship to a phage's fighter jets. A phage measures 225 nanometers (nm) at most at its longest end-to-end distance; a typical bacterial cell volume is 1,300 times the volume of a phage.

Microbiologists have revived Felix d'Herelle's idea of a century ago by designing phages to enter bacteria and inactivate bacterial repair kits or shut down antibiotic pumps. This method has already been tried in humans to correct genetic diseases in the new science of gene therapy. In gene therapy, molecular biologists engineer viruses that infect humans to contain a specific gene that will repair a defect in human DNA. They inactivate the virus so that it cannot cause disease but can still infect the human host. When the engineered virus takes over the cell's DNA replication, it inserts the new gene into the defective DNA.

Phages built to deliver antibiotics or foil the defenses of resistant bacteria have not been tried outside laboratory trials. But because of the constant evolution of bacteria to avoid harm from drugs, biology must stay abreast with new weapons of its own.

## Bacteria share their DNA

Gene transfer confers on bacteria the capability to accept helpful genes from other microbes. In eukaryotes from algae all the way up to humans, gene transfer occurs by one mechanism, the fusion of gametes. One gamete from a female and one from a male creates a zygote that carries the DNA from both parents. Bacteria and archaea have three major routes whereby they exchange genes: transformation, transduction, and conjugation. All of these methods are called horizontal gene transfer because they occur between two or more adult cells rather than the standard sharing of genes by producing daughter cells.

Transformation occurs when bacteria take in DNA directly from the environment. The DNA may be either the molecule from the nucleoid or a plasmid. In either case, the DNA dissolves in an aqueous environment when a cell dies and lyses. A live bacterial cell encountering the DNA in its habitat may attach to the molecule and use an enzyme to unravel the large polymer. DNA is a double-stranded structure resembling a ladder. The enzyme cuts the ladder's rungs to separate the DNA in half. One half degrades, but the cell pulls the other half inside where it will incorporate it into its own DNA.

Transduction occurs when a bacteriophage infects a bacterial cell and brings DNA from another microbe with it. If the phage commandeers the cell's DNA replication steps but does not kill the bacterium, the bacterial cell makes new progeny containing some of the foreign DNA. New bacteria never before seen in nature begin growing.

When plasmids transfer between cells, they do so by conjugation. Conjugation has been called the bacterial version of sexual reproduction because two cells physically connect with one another by a tube called a sex pilus. After DNA has moved through the pilus from the first cell to the second, the pilus breaks. As a result of conjugation, the receptor cell incorporates new genes into its existing DNA. When the cell divides, the daughter cells and each successive generation can carry these genes.

Gene transfer in bacteria has its most profound effects in allowing antibiotic-resistance genes to move through a population of bacteria. The bacteria need not be closely related as long as they can use one of the three methods described above for passing DNA back and forth. Since plasmids have been shown to carry multiple genes for antibiotic resistance, plasmid transfer may be a major route for the expansion of antibiotic resistance in the past few decades. Biologists have not answered all their questions on the evolution of gene transfer in bacteria, but there can be no question of the advantages these systems give to bacteria.

## The opportunists

Hospitals act as hot spots for antibiotic-resistant bacterial infections because hospital settings have high antibiotic use and a patient community weakened by disease, trauma, or surgery. These circumstances open the opportunities for bacterial infection. Nosocomial infections are infections picked up in hospitals. Many of these infections could well come from doctors, nurses, technicians, and other hospital staff who do not wash their hands properly between patient visits. Secret observations of hospital staff have revealed that healthcare professionals wash their hands properly only slightly more often than the general public of which less than 50 percent wash properly. Most of these poor habits (not enough time washing hands, not enough soap, no soap, or no hand wash at all) occurred in the public restroom! Most hospitals now have resident bacteria in proportions found nowhere else in society, and these nosocomial populations have a high incidence of multidrug resistance. No wonder that people believe that any bacterium is a dangerous bacterium. This thinking spawned not only antibiotic misuse, but a similar overuse of disinfectants and other antimicrobial products.

Medical microbiologist Stuart Levy has warned that overzealous cleaning with disinfectants merely increases the opportunity for bacteria to develop resistance. Might disinfectant- and antibiotic-resistant superbugs share their best defenses with each other by exchanging genes? Such sharing seems implausible because the chemicals in cleaning products (bleach, quaternary ammonium compounds) differ from large antibiotic molecules. Yet bacteria eject these chemicals much the same way they expel antibiotics: They use a pumplike mechanism. The term "pump" can be misleading. Bacterial antibiotic efflux pumps use transporters inside the cell. When an antibiotic enters the cell through a receptive pore in the bacterium's outer membrane, the transporter moves toward the antibiotic and locks onto it. A bacterial protein (called a fusion protein) then recognizes the transporter now reconfigured by the antibiotic and swiftly carries the complex out through another pore. As long as bacteria have the nutrients needed to build transporters and fusion proteins, they can resist antibiotics by excreting them. Because the transporter must recognize all or part of the antibiotic for this system to work, chemists try to construct unique antibiotics, and biologists seek new natural substances that will throw a monkey wrench into the antibiotic efflux pump. If molecular biologists discover that the chemical pump and the antibiotic pump are one and the same, a new super-superbug may be around the corner, able to resist disinfectants as well as it resists antibiotics. No one yet knows which side will win the race to perfect resistance or a perfect drug.

Surely the rise in antibiotic resistance has made a difference to the bacteria that have always lived in harmony with their host. When the body's good bacteria cause infection, they do so because circumstances change to invite them in. These circumstances usually have to do with a weakened or immature immune system, mainly in groups of people considered "high-risk" individuals:

- Chronic, debilitating disease
- Drug or alcohol abuse
- Poor nutrition
- Pregnancy
- Old age
- Young age (infants and children under 12 years)
- HIV/AIDS
- Organ transplantation
- Cancer chemotherapy or radiation.

Each of the stressors listed here increases the dangerous cycle of antibiotic-resistance causing infection that requires antibiotics, leading to more resistance. One of the prevalent bacteria on the body, *Staphylococcus aureus*, has already become one of the most multidrug-resistant microbes known. Because *S. aureus* is both a health risk and a prominent member of the body's normal flora, good personal hygiene usually trumps antibiotics, disinfectants, and other weapons from the antimicrobial armory (see Figure 3.2).

Drug companies have for the past decade introduced fewer and fewer new antibiotics. Because "all the easy antibiotics have been discovered," research into new natural or synthesized compounds has grown more difficult and more expensive. Companies that once led in antibiotic production have now decreased the money they spend on new antibiotic research. The combination of skyrocketing research costs and patents that limit the profit-earning future of drugs has left doctors with a shrinking armamentarium against infectious disease.

Entrepreneurs have tried colloidal silver, copper, zinc, magnesium, medicinal herbs (cloves, echinacea, garlic, oregano, turmeric, and thyme), citrus oils, tea tree extracts, and grapefruit seed extract. I have tested most of these substances on laboratory cultures, and they do possess antibacterial activity. But inhibiting bacteria in a laboratory is much easier than stopping bacteria in nature or in the body. In a laboratory, bacteria are at their most vulnerable to damage because antibiotics work best on rapidly multiplying cells. In nature, bacteria turn on defensive mechanisms and slow their growth. Both actions take away some of the power of antimicrobials.

Figure 3.2 Court at No. 24 Baxter Street, ca. 1890. Photographer Jacob Riis captured life in one of New York City's tenement slums. Similar living conditions exist today worldwide. Poor nutrition and faulty hygiene have contributed to germ transmission throughout history. (Courtesy of Museum of the City of New York, Jacob A. Riis Collection)

A new generation of antibiotics may yet emerge. If they do, they will probably come from the ocean. In the past decade scientists have recovered marine bacteria, algae, sponges, coral, and microscopic invertebrates that produce novel antibiotics. The new marine antibiotics might soon replace current antibiotics that are losing the battle against *Staph* infections, gonorrhea, strep, tuberculosis, and nosocomial infections.

### The history of medicine

2000 BCE—Here, eat this root.

1000 CE—That root is heathen. Here, say this prayer.

1850 CE—That prayer is superstition. Here, drink this potion.

1920 CE—That potion is snake oil. Here, swallow this pill.

1945 CE—That pill is ineffective. Here, take this penicillin.

1955 CE—Oops...Bugs mutated. Here, take this tetracycline.

1960-1999 CE—Thirty-nine more "oops." Here, take this more powerful antibiotic.

2000 CE—The bugs have won! Here, eat this root.

—Anonymous (2000)

# 4

# Bacteria in popular culture

Bacteria and viruses are silent, invisible, and multiply inside the body. Sometimes they mutate; sometimes they kill. No one could blame a novelist for making microbes into antagonists for a hero to overcome. Bacteria have for decades infiltrated popular culture, and the arts offer a surprising number of lessons on disease as well as Earth ecology. As important, the arts have communicated people's fears and conceptions of bacteria. Misconceptions about bacteria in movies and novels reveal how people view germs. The perceptions of bacteria give insight into the effects bacteria have had on society and events in the past.

Popular culture, regardless of the century, has understandably made more of deadly pathogens and given less credit to the environmental microbes that make the planet livable. The exaggerations and falsehoods regarding pathogens in the arts enlighten us to the perceptions of bacteria that have persisted through the years.

## Bacteria and art

Europe's Black Plague influenced art and mirrored changing attitudes toward disease and death. Early 14th-century paintings before the plague depicted serene country life, the hunt, and the upper classes. The church often influenced the work—Heaven and Hell received almost equal focus—but artists seldom made death appear violent or cruel. When the Black Death tightened its grip on upper and working classes alike, artwork reflected the somber mood. As the plague and its toll continued with no end in sight, European artists conveyed only the tragic and painful outcomes society faced. Heaven and Hell no longer shared equal billing; the jaws of Hell seemed to gape everywhere. Europe's 14th century painting in fact displays a multitude of deathbed scenes.

Hidden behind the images was an oppressive darkness that *Y. pestis* brought to the continent. The plague microbe had developed no special traits that allowed it to emerge with regularity and unhindered in Europe for five centuries. Crowded cities, poverty, misinformation, and perhaps too much faith in a powerless clergy and medical profession made the plague into the scourge that changed history. These same factors, more or less, exist today.

The Black Death also affected artists' lives in an unexpected way. Because the disease interrupted invasions on Europe by Barbarian tribes, small and large European towns had time between epidemics to develop creative pursuits. Artists, skilled artisans, and architects became proficient in their crafts, and they rose to professional stature and enhanced level of respect in society.

No one at the time of the Black Death had a notion as to its cause. Antoni van Leeuwenhoek would not view bacteria in a microscope for another three centuries. Historians gleaned from artwork the misery that *Y. pestis* caused. Paintings showed pale, weak subjects fallen to the ground where they had stood. Often crowds of the sick and dying shuffled past the corpses. Almost every account of the plagues from Justinian's through the Great Plague of London in 1665 described bodies piling up in the streets. These accounts and the art of the period captured not only the despair of the surviving but also their challenges. Paintings and writings described townspeople hauling bodies to distant funeral pyres by handling the dead with long sticks or poles, trying to avoid too-close contact with the contagion.

## Bacteria in the performing arts

A familiar rhyme thought to have originated during London's Great Plague in 1665 has developed different versions in various languages and cultures over time, but all convey the same message:

> Ring around the rosey,
> A pocketful of posies.
> Ashes, ashes.
> We all fall down!

A microbiologist living in the Middle Ages but armed with today's knowledge of bubonic plague might revise the rhyme to a less lyrical:

Red rash encircling the bulbous swelling of the skin,

A supply of medicinal herbs.

Burn the deceased in funeral pyres.

We all die from the plague sooner or later.

The plague struck down its victims within hours. A healthy person infected with *Y. pestis* in the morning could be dead by nightfall. But plague epidemics grew less frequent between the 15th and 19th centuries—the reason for this has not been fully explained. As the plague disappeared, another disease haunted society, and thus entered the arts. Tuberculosis (TB), known as consumption into the early 1900s, is thought to be humanity's oldest disease. The lengthy and debilitating illness causes a slow decline in many of the people who do not receive treatment. *M. tuberculosis* takes 24 hours to divide in two, and TB thus develops very slowly in an infected person.

*M. tuberculosis*'s curved rods reach no more than 4 µm long and 0.3 µm wide. These stringy bacteria travel through the air in moisture droplets expelled by the cough of an infected person. The droplets called bioaerosols can drift in the air for several feet before being inhaled by a new host. Once inhaled, as few as five *M. tuberculosis* cells begin an infection by infiltrating the air sacs, or alveoli, of the lungs. The host's immune system responds to the presence of the foreign entity by sending macrophage cells to the site of infection. The macrophages engulf *M. tuberculosis* as they do all other foreign matter with the intent of decomposing it. But macrophages cannot kill *M. tuberculosis*. Some of the bacteria hide inside the macrophage and ride with it through the lymph system to other organs. Other *M. tuberculosis* cells stay in the lungs and multiply. The intensifying infection prompts the immune system to double its efforts, and so an increased inflammatory reaction develops in an effort to kill the infection. As a result, the body's immune system causes more harm than the bacterium.

The immune system's fruitless attempt to fight TB contributes to the disease's severity. Most bacteria would be destroyed by a healthy person's immune response, but *M. tuberculosis* gathers the immune cells around it until a mass, a tubercle, forms in the lung. Each small cluster of *M. tuberculosis* cells builds numerous tubercles throughout the lungs. Lymph fluids begin to accumulate in the organ, and the inflammation creates lesions in the tissue. The infected person develops TB's telltale chronic cough.

A slow decline of an afflicted character helped create storylines for *La Traviata* and *La Bohème*. Departures of hefty opera divas to a disease that actually leaves its victims weak and emaciated never got in the way of the production. Earlier in the 18th century, English physician Benjamin Marten had made an astute observation and proposed that "wonderfully minute living creatures" might be the cause of consumption. In his writings, Marten discussed the potential risk of healthy individuals living in close contact with the infected. These ideas were ahead of their time. A few doctors advised that the infected refrain from close contact with others, but families often rejected this idea as cruel punishment rather than a preventative. Into the 1940s, medicine still had no reliable treatments or accepted preventions for TB.

For two decades, the health community promoted sanatoria for the seclusion of TB patients and recovery without transmitting it to others. Patients spent several months to a year away from their families. (Doctors prescribed complete rest, even limiting patients' bathroom breaks to one per day.) Imagine the fertile ground dramatists mined by sending a character far from family, lovers, or creditors! In 1945, screenwriter Dudley Nichols banished Sister Mary Benedict played by Ingrid Bergman to a TB sanatorium in the melodramatic ending of *The Bells of St. Mary's*.

In the overall assessment of how diseases move through populations, TB behaves in complete contrast to bubonic plague. The highly virulent plague bacteria kill victims swiftly. The worst of history's plague epidemics have wiped out populations nearly en masse, forcing the pathogen to retreat to its rodent reservoir for a while. TB's slow progression through a population and long course enables it to remain in a population longer than acute diseases. TB does not always kill but merely incapacitates the host, which further helps it infiltrate entire communities.

The sanatoria depicted as unjust segregation of the sick was in truth the best way to stop the transmission of infectious disease and remains so today. TB is a social disease. Close interaction between people, crowded living and workspaces, and frequent movement of the infected to new areas help TB persist in society. Society preferred, not for the first time, to equate social disease with destitution, lack of education, and low social standing. The public had a hard time shaking the belief that TB was somehow a person's fault. This philosophy continues today with other bacterial diseases and viruses. Despite all of the technological advances microbiology has made, many still view infection in a spiritual sense rather than as a biological reality.

Aside from retreating to a TB sanatorium, many natives of cold, crowded cities in the east sought a warm place to recuperate for a year or more from a debilitating disease. California's movie industry grew in part due to an increased population lured by "a climate that makes the sick well and the strong more vigorous," as Chamber of Commerce brochures claimed early in the 1900s. Families that had been affected by TB or wished to avoid it made cross-continent moves to sun-baked southern California.

TB took lives from the arts as it had from every other faction of society. Most of the famous who succumbed to TB, shown in Table 4.1, died young, and this list illustrates the pervasiveness of TB into the 20th century.

**Table 4.1** Famous TB victims

| Name | Date | Contribution |
|---|---|---|
| Alexander Pope | 1744 | British poet and satirist (age 56) |
| John Keats | 1821 | British Romantic poet, wrote "Ode to a Nightingale" (1819) (age 26) |
| Percy Bysshe Shelley | 1822 | British Romantic poet, wrote "Prometheus Unbound" (1820) (age 30) |
| Johann Wolfang von Goethe | 1832 | German author of *Faust* (1808) (age 83) |
| Emily Brontë | 1848 | British author of *Wuthering Heights* (1847) (age 30) |
| Frédéric Chopin | 1849 | Polish pianist and composer (age 39) |
| Edgar Allan Poe | 1849 | American poet and short-story writer, wrote "Murders in the Rue Morgue" (1841) (age 40) |
| Charlotte Brontë | 1855 | British author of *Jane Eyre* (1847) (age 39) |
| Elizabeth Barrett Browning | 1861 | British Victorian poet, wrote "Sonnets from the Portuguese" (1850) (age 55) |

**Table 4.1** Famous TB victims

| Name | Date | Contribution |
|---|---|---|
| Henry David Thoreau | 1862 | American writer and philosopher, author of *Walden* (1854) (age 45) |
| Stephen Foster | 1864 | American composer of "My Old Kentucky Home" (1853) (age 38) |
| Fyodor Dostoyevsky | 1881 | Russian author of *The Brothers Karamazov* (1880) (age 60) |
| Robert Louis Stevenson | 1894 | Scottish author of *Strange Case of Dr. Jekyll and Mr. Hyde* (1886) (age 44) |
| Anton Chekhov | 1904 | Russian playwright and short-story writer, wrote *The Seagull* (1896) (age 44) |
| Franz Kafka | 1924 | Austria-Hungarian author of *The Metamorphosis* (1915) (age 41) |
| D. H. Lawrence | 1930 | British author of *Lady Chatterley's Lover* (1928) (age 45) |
| Thomas Wolfe | 1938 | American author of *Look Homeward, Angel* (1929) (age 38) |
| George Orwell | 1950 | British author of *Nineteen Eighty-Four* (1949) (age 47) |
| Vivien Leigh | 1967 | British actress played Scarlett O'Hara in "Gone with the Wind" (1939) (age 54) |
| Igor Stravinsky | 1971 | Russian pianist and composer (age 89) |

Outside the arts, King Edward VI (age 16), Doc Holliday (age 36), and Eleanor Roosevelt (age 78) succumbed to the disease as did Rene Laennec (age 45), the inventor of the stethoscope. Some historians have implied that George Washington died of TB—the disease had claimed his brother Lawrence—but definite proof has never been found. The Father of Our Country had been sickly most of his life and may have suffered two bouts of TB. On December 14, 1799, Washington died of what his doctor called a case of "inflammatory quinsy" in the respiratory tract. Generations of medical researchers have puzzled over the cause of Washington's death. No argument exists on the death of Henry Livingston Trudeau, the proponent of sanatoria in the United States. Trudeau's repeated exposure to the people he tried to save likely caused his infection and death at age 67.

Modern poet Dylan Thomas did not die of TB but, according to medical historian H. D. Chalke, he became so obsessed with TB he might as well have contracted it. Thomas's repeated references to looming death are thought to provide evidence of the poet's fear of the disease.

## Friends and enemies

Authors have used bacterial diseases as metaphors for various states of the human spirit and body. The Brontës, Jane Austen, and Charles Dickens alluded to TB in their novels, especially when guiding a character into imminent suffering, as did John Steinbeck in the somber plots of *The Grapes of Wrath* in 1939 and its 1938 prelude *Their Blood Is Story*.

In the 1800s and 1900s, the waterborne disease cholera ranked second only to TB among infectious diseases in causing death. In the 1912 novella *Death in Venice*, Thomas Mann killed off his protagonist, the aging artist Gustav von Aschenbach, with cholera to save him from the agony of a sexual obsession. W. Somerset Maugham's *The Painted Veil* (1925) and Gabriel García Márquez's *Love in the Time of Cholera* (1985) also used the swift and deadly disease as a vehicle for advancing their stories. Cholera, TB, or any of the diseases for which cure is elusive have contributed to metaphors on the inevitability of death, infirmity, loss, and the emotions that will always drive literature, music, and the visual arts.

In 1938, a radio drama broadcast from New York City's Mercury Theatre created a rare hero's role for bacteria. At eight o'clock on Halloween Eve, actor Orson Welles stepped to the microphone. For the next hour he reported to an increasingly frantic radio audience the takeover by Martians of the world's major cities. Scientists, the military, and negotiators failed to stop the invasion. Humans, it seemed, were about to be wiped from the Earth. Near the final minute of the broadcast the protagonist found that the Martians had fallen "stark and silent" with vultures picking at the remains. Microbiologists listening in that night knew that the probable hero to save humanity would be "the putrefactive and disease bacteria against which [the Martians'] systems were unprepared...slain, after all man's defenses had failed, by the humblest thing that God in His wisdom put upon this earth." Welles's description of bacteria as putrefactive and disease-causing may have been a bit ungrateful considering they had just saved the planet, but "The War of the Worlds" taught the basic truths of bacteria: Any bacterium can turn deadly in hosts with weakened immune systems.

Since the 1938 broadcast, microbiologists have learned much more about the resiliency of Earth's collective population of bacteria. Bacteria that withstand, heat, cold, radioactivity, intense pressure, desert-dry conditions, ultraviolet light, chemicals, and lack of oxygen have been isolated, studied, and put to productive use. With antibiotic-resistance having spread to the majority of humanity's worst pathogens, bacteria may someday depopulate the planet in the same way they eliminated Orson Welles's Martians.

I have always appreciated that the bacteria in "The War of the Worlds" saved the day. I'm doubly happy that Welles did not commit the frequent error of confusing bacteria with viruses.

In 1987, novelist Michael Crichton returned bacteria to their more standard role as indestructible enemies of man. With the time-honored gimmick of a mutant organism arriving from space to destroy humanity, *The Andromeda Strain* gave a detailed and mostly correct view into microbiology few people know about: the techniques used in cultivating life's deadliest pathogens.

Crichton accurately described the intensive precautions microbiologists take when working with the world's most virulent pathogens. These labs are called Biosafety Level 4, or BSL-4 laboratories. BSL-4 labs contain special air circulation and filtration systems, multiple air-locks to prevent the escape of airborne microbes, the use of protective clothing, and decontamination measures before anyone can enter or exit the lab. The author suggested that "disinfectants" such as ultraviolet or infrared light, ultrasonic waves, or flash-heating would sterilize the fictional characters' bodies. In fact, these methods harm a person more than they hurt bacteria; the human body cannot be sterilized. Disinfectants work only on inanimate objects. Antiseptics, not disinfectants, remove some but not all bacteria from the skin.

Despite the book's minor mistakes, *The Andromeda Strain* conveyed many excellent points about the lifestyle of bacteria. Crichton accurately described extremophiles and bacterial spores. He included an example of Koch's postulates in which a microbe can be proven to cause specific disease by taking it from a sick organism, injecting into a healthy individual, and re-creating the disease.

Crichton's fictional strain grew only on carbon dioxide, oxygen, and sunlight, and required a very narrow pH range, meaning the relative amounts of acid and base in its environment. The Andromeda Strain also derived nutrients by eating the rubber gaskets of enclosures the novel's scientists had hoped would confine the pathogen. Crichton had described what microbiologists call a photoautotroph, which is a bacterium that exists only on sunlight for energy, carbon dioxide for carbon, and very few other nutrients. In the evolution of life on Earth, photoautotrophs generated the first traces of oxygen in the atmosphere. Other photosynthetic bacteria followed, and they added more oxygen to the atmosphere, paving the way for the evolution of invertebrates, fish, mammals, and all other oxygen-requiring organisms.

Rubber-eating bacteria are not unusual. At least 100 different rubber-degrading bacteria have been identified, and many more unidentified strains exist. Both bacteria and fungi degrade the five-carbon, eight-hydrogen isoprene units of natural rubber, such as the rubber in latex gloves. In 2008, Mohit Gupta of Drexel University College of Medicine made the disturbing discovery of a *Gordonia polyisoprenivorans*-caused pneumonia in a hospital patient. This bacterium normally grows in the stagnant water inside discarded tires, slowly eating away at the hard, black rubber. Perhaps the massive mountains of refuse tires throughout the United States will inspire a new science fiction thriller.

The scientists in *The Andromeda Strain* never found their magic bullet against the invader. The pathogen disappeared as many do by mutating to a less virulent form and destroying too many of its hosts. The book's outcome should sound familiar: Medicine never defeated the bubonic plague because it defeated itself.

## Do bacteria devour art?

Who would guess that studying ancient artwork offers an excellent opportunity to learn about bacterial metabolism? Bacteria degrade art and historic treasures in the same way they decompose organic matter. Bacterial enzymes lipase and protease break down fats and peptides, respectively, in pigments and a variety of carbohydrate and fiber-degrading enzymes attack the canvas and wood. These are the same enzymes animals use for digesting food. Different, more specialized bacteria get energy from chemical reactions involving inorganic salts. All of these microbial actions contribute to the slow decomposition of the world's greatest works of art, principally because of bacteria acting in concert within a community (see Figure 4.1).

Figure 4.1 Biofilm bugs. (Courtesy of Center for Biofilm Engineering, Montana State University)

The decomposition of artwork is one small piece of bacterial cycling of nutrients on Earth. Bacteria circulate the Earth's elements through the atmosphere, water, soil, and plant and animal life in processes called nutrient or biogeochemical cycles. These cycles take place in the seas, forests, and mountains. Nutrient cycling also occurs when bacteria decompose rubber, plastic water bottles, paint, and countless manmade items that were once believed to be indestructible. Bacteria corrode metal, stone, marble, and concrete, and they degrade paint, paper, canvas, leather, pigments, and wood. Chemical reactions driven by bacteria weaken modern infrastructure such as bridges, roadways, and oil tankers. In exactly the same way, bacteria have been steadily digesting the components found in art, whether these are made of metal, fiber, hide, or pigments.

Copper is one of the oldest metals used in civilization. In the Bronze Age (3000-1300 BCE) craftsmen took advantage of the metal's malleability to incorporate it into the alloys brass and bronze for tools, weapons, domestic items (bowls, plates, goblets, and so on), and jewelry. Microbiologists have learned just in the past several years that sulfate-reducing bacteria have been corroding bronzes such as Etruscan relics dating to the ninth century BCE. When a bacterium is said to "reduce" an element, it means a bacterial enzyme adds electrons to the element. In metal corrosion, sulfate-reducing bacteria convert sulfate, a sulfur atom with oxygens attached, to the element sulfur. When a biofilm forms on relics containing iron, the anaerobic bacteria at the base of the film convert sulfur to pyrite, an iron atom attached to two sulfurs. Figure 4.2 illustrates the substantial biofilm growth that metal structures can support.

Figure 4.2 Biofilm corrosion. This pipe has been almost completely occluded by biofilm, which has dried and hardened. Few technologies exist for removing biofilm from living or nonliving surfaces. (Courtesy of Center for Biofilm Engineering, Montana State University)

Sulfate-reducing *Desulfovibrio* and the iron-oxidizing *Leptothrix* work in concert to corrode iron; they are sometimes nicknamed "iron-eating" bacteria. *Leptothrix* takes electrons away from iron atoms at the metal's surface, and *Desulfovibrio*, hovering a few μm nearby, accepts the excess electrons. Even though iron corrodes when exposed to the air, the biofilm actually creates tiny anaerobic nooks called microenvironments in which these reactions take place. Sergei Winogradsky discovered the general steps in iron-sulfur metabolism between 1885 and 1889.

Bacterial deterioration of metal takes place every day 12,850 feet at the bottom of the Atlantic. The *H.M.S. Titanic* has withstood the incredible hydrostatic pressures exerted on it for a century. The oxygenless deep waters have also allowed the ship to resist rusting. Yet the *Titanic* supports a ghostly collection of rusticles. These appendages from several inches to a few feet long and numbering in the thousands hang from almost every part of the ship. Some are as fragile as tissue; others hold their shape when research vessels pull them to the surface. The rusticles demonstrate that the main cause for the *Titanic*'s inexorable return to the Earth is bacteria.

The rusticles contain a mixture of bacteria able to thrive in the cold and deep where the *Titanic*'s wreck settled on April 15, 1912. The bacteria remove 0.3 gram of iron from every square centimeter of the ship daily. The loss of iron causes about 300 kilograms of steel to detach from the wreck each day. Anaerobic "iron-eating" bacteria have been taking apart the *Titanic* one iron atom at a time and may cause the hull to cave in on itself within 100 years, perhaps as little as 40 years from now. The ship's organic material, mostly wood paneling and fixtures, serve as the main nutrient source for the bacteria, but as the metals corrode more organic matter may be exposed. As a result, the deterioration of the *Titanic* will accelerate.

Stone and concrete undergo similar aboveground weathering. The corrosion of ancient Greek and Roman stone statues illustrate bacteria's role in the slowest of all biogeochemical cycles, the rock cycle or sediment cycle. Unlike nitrogen, which can migrate from soil to the atmosphere and return to plants within a day, the rock cycle takes eons to complete. Bacteria begin by weakening the stone through corrosion. Small pieces break off from the mass and make their way downhill with erosion. Erosion carries the matter to bodies of water where it sinks and becomes part of sediment. Sediments, especially under the ocean, compact under intense pressure. As tectonic plates shift, this sediment becomes part of metamorphic rock in the Earth's mantle and slowly pushes upward to the planet's surface. Some metamorphic rock sinks into the Earth's interior where the planet's molten core heats the sediment and turns it into magma. Magma rushes to the surface all at once in volcanoes. New rock that has either migrated up to the Earth's surface gradually or exploded toward the surface from a volcano becomes available for bacteria to again begin degrading.

Molecular methods have now been applied to studies of how bacteria affect not only rock, but also prehistoric paintings in moist caves. The 20,000-year-old cave paintings in Altamira, Spain, and Lascaux, France, may be succumbing to the combined activities of bacteria that degrade the dyes as well as the underlying stone. Many of the offending bacteria have not yet been identified, but microbiologists have noticed that *Actinobacteria* often dominate the microbial populations in the Spanish and French caves. *Actinobacteria* build tentacles called filaments that grow into the pores of rock surfaces, allowing the bacterial damage to occur in the rock's subsurface. Molecular analyses of the cave paintings have also uncovered aerobes (*Pseudomonas*) and anaerobes (*Thiovulum*), sulfate- (*Desulfovibrio*) and iron-users (*Shewanella*), bacteria that use a wide variety of nutrients (*Clostridium*), and species with narrow nutrient requirements (*Thiobacillus*). Lascaux's 600 paintings made of a mixture of mineral pigments and animal fat offers a banquet for the bacteria of the caves.

Art galleries have almost as difficult a time protecting treasures from bacterial decay, despite humidity and temperature-controlled environments. Fungi and the funguslike bacterium *Actinomyces* extend filaments into paintings' surfaces to cause physical destruction. Other microbes chemically decompose the pigments. Polymerase chain reaction (PCR) technology, invented in the 1980s, has enabled microbiologists to multiply bits of bacterial DNA recovered from painted frescoes on castles in Austria, Germany, and France to study the types and proportions of the bacteria. The analyses have so far revealed the presence of *Clostridium*, *Frankia*, and *Halomonas* on the ceiling painting in Castle Herberstein in Styria, Austria. Each genus contributes its own mode of destruction on the painting:

- ***Clostridium***—A spore-forming anaerobe that grows well on a wide variety of chemicals
- ***Frankia***—A spore-forming genus that builds long, branching filaments that penetrate surfaces
- ***Halomonas***—A versatile halophile that lives with or without oxygen and can degrade alcohols, acids, and organic solvents

The constant streams of tourists who visit the world's great works of art accelerate decay. Human bodies and breath change the temperature and humidity in galleries and even in caves containing ancient wall paintings. The Lascaux caves had been in good condition when discovered in 1940. The rapid decay of the cave paintings that took place after people started visiting then led to the closure of Lascaux in 1965 to prevent further deterioration.

On stone exposed to the weather, biofilms and cyanobacteria each contribute in their own way to the deterioration of statues, buildings, and headstones. In some instances, bacterial growth on historical structures presents no more than a cosmetic problem due to the discoloration of stone by bacterial pigments. In other cases, acids produced by bacteria in biofilms degrade the stone's calcium carbonate as they degrade tooth enamel in dental caries. The actions of fungi, biofilm bacteria, and free bacteria with their different types of metabolism hasten the decomposition of historical structures and have already decimated most concrete structures, not only from ancient periods but since the 20th century.

Lichens form greenish black stains on old stone structures. Lichen is a living entity made from a cooperative relationship between a fungus and a bacterium or an alga. Among bacteria, cyanobacteria are by far the most common to form associations with fungi. The photosynthesis performed by cyanobacteria helps keep the lichen alive but also may aid the growth of other bacteria by providing organic nutrients. The limestone surfaces of the Mayan ruins at Chichen-Itza support bacteria and other microbes. The bacterial populations become denser and more diverse on the sun-bathed stones and less dense and varied on surfaces inside temples and corridors.

Microbiologists have taken a unique tack for preventing the further deterioration of art by using other bacteria to combat the effects of art-degrading species. The first task involves cleaning the objects to remove the heavy detritus that has accumulated over centuries. Specific bacteria remove crusts of sulfates and nitrates, animal-based glues, and the remains of molds and insects. Injecting nutrients into the pores of art's materials might allow bacteria to grow and form crystals that would block future infiltration of the porous surfaces. Similar bacteria may be applied to clean the microscopic crevices of a piece. Researcher Giancarlo Ranalli, working in Pesche, Italy, has become an expert on using bacteria to clean art. He has applied bacteria-filled poultices to clean the marble of Michelangelo's Pieta Rondanini, and in 2007 his team reported a comparison between bacteria and a cleaning mixture of ammonium carbonate, detergent, and an abrasive applied to marble surfaces in Milan Cathedral. *Desulfovibrio vulgaris* cleaned the marble without removing the material's patina while the cleaning mixture removed less dirt and left behind a precipitate. Ranalli's team has since put *Pseudomonas stutzeri* to work digesting glue from protective shrouds that had covered frescoes in the Pisa Cemetery for more than 20 years. In this case, *P. stutzeri*'s unique protein-degrading enzymes released the fabric without destroying the fresco underneath.

Conservators of Europe's galleries have been hesitant to let a microbiologist spread bacteria all over their valuable works of art. Although bacteria like Ranalli's have worked on stone, the same method does not have a long track record on gallery art. Cleaning a 300-year-old painting differs from using bacteria to dissolve crud from inside restaurant grease traps, septic tanks, and wastewater holding tanks in navy ships, all current uses for *Bacillus*, *Pseudomonas*, and other bacteria. Art-cleaning bacteria will require fine-tuning to ensure they digest unwanted dirt but leave the art's components in good condition.

Biotechnology might help in the new field of bacteria-based art refurbishing. Bacteria can be engineered to excrete antibiotics targeted to art-loving bacteria. Genetically modified organisms (GMOs) might also someday act on specific components in art buildup, and then stop due to a shut-off gene that activates when the target compounds are gone. In the meanwhile, bacteria gnaw on the Roman Coliseum and perhaps the Mona Lisa.

# 5

## An entire industry from a single cell

The biotechnology industry arrived in the late 1970s when entrepreneur-biologists began harnessing microbes for profit. A new company, Genentech, first entered the commercial market in 1977 with the peptide somatostatin, made by *E. coli* engineered to carry genes that encoded for this growth-modulating hormone. Prior to these *E. coli* fermentations, somatostatin came only from cattle after slaughter.

The first success in moving genes from one organism into a different, unrelated organism occurred in 1972 in Paul Berg's laboratory at Stanford University. Berg composed a hybrid DNA from the DNA molecules extracted from two different viruses. The next year Herbert Boyer and Stanley Cohen further stretched the boundaries of gene transfer by putting genes from a toad into *E. coli*. Most important, successive generations of the engineered *E. coli* retained the new gene and reproduced it whenever they made new copies of *E. coli* genes. Boyer and Cohen had developed recombinant DNA, and as a result, the world had its first human-made GMO.

Some biotechnologists have taken a broad view of when their science began, citing the first use of bacteria or yeasts to benefit humans. Using this criterion, biotech began in 6000 BCE when people first brewed beverages using yeast fermentations. For practical purposes, the science of manipulating microbial, plant, and animal genes emerged when scientists first cleaved DNA and then inserted a gene from an unrelated organism into it. The biotech industry commenced when companies made the first commercial products from recombinant DNA by growing large volumes of GMOs.

Berg, Boyer, and Cohen would not have initiated the new science of genetic engineering without prior individual accomplishments in genetics. Walther Flemming, in 1869, collected a sticky substance from eukaryotic cells he called chromatin, later to be identified along with associated proteins as the chromosome. In most bacteria, the chromosome is a single DNA molecule packed into a dense area of the cell (called DNA packing). Bacteria do not contain the proteins, called histones, which eukaryotes use for keeping the large DNA molecule organized. Eukaryotes carry from one to several chromosomes. The eukaryotic chromosomes plus DNA located in mitochondria collectively make up the organism's genome. In bacteria, the genome consists of the DNA plus plasmids.

In the early 1900s, Columbia University geneticist Thomas Hunt Morgan used *Drosophila* fruit flies to demonstrate that the chromosome, in other words DNA, carried an organism's genes. Less than 50 years later, American James Watson and British Francis Crick, who were both molecular biologists, described the structure of the DNA molecule.

DNA structure resembles a ladder that has been twisted into a helix. The long backbones, or strands, consist of the sugar deoxyribose, each holding a phosphate group (one phosphorus connected to four oxygens) that extends away from the ladder. Deoxyribose also holds a nitrogen-containing base on the opposite side that holds the phosphate group. Each base points inward so that different bases from each strand and complementary in structure connect by a chemical bond. These bonds, called hydrogen bonds, hold atoms together by weak connections compared with other types of chemical bonds.

Nature uses only four bases in DNA to serve as a type of alphabet. These bases are adenine, thymine, cytosine, and guanine, which biologists abbreviate to A, T, C, and G, respectively. The sequence of bases in DNA determines the makeup of genes, which are short segments of bases. The exact sequences of A, T, C, or G in each living organism hold all of the genetic information that defines the organism's species and also makes every individual unique. No two DNA compositions are identical.

Paul Berg and the other leading molecular biologists first created hybrid DNA by cutting the strands with an enzyme called restriction endonuclease. (Restriction endonucleases evolved in bacteria for the purpose of destroying foreign DNA brought into the cell by an invading phage.) The break in the DNA molecule served as a place to insert one or more genes from another organism.

An alphabet composed of only four letters does not seem adequate for carrying all the heredity of every organism on Earth. Nature solved this potential problem by requiring that each sequence of three bases serve as the main unit of genetic information, called the genetic code. The base triplet constitutes a codon, and each codon translates to one of nature's amino acids, which act as the building blocks of all proteins regardless of whether the proteins belong to animals, plants, or microbes. Only 20 different amino acids go into nature's proteins that vary in length from about 100 to more than 10,000 amino acids. The three-letter codons increase nature's capacity to put all of its information into genes made of no more than four letters. The varying lengths of proteins further expand the possibilities for defining everything in nature from a simple microbe to a human. The genetic code furthermore defines every being that once lived but has gone extinct.

Imagine if only one base were to encode for one amino acid. Proteins would not be able to contain more than four amino acids. A codon composed of two bases could hold a maximum number of amino acids of $4^2$, or16. By adding one more base, the maximum number of amino acids that the alphabet could define would be $4^3$, equal to 64. DNA's triplet codons can thus identify all of the essential amino acids with several codons to spare. Because nature tries to do things in the simplest way possible, it has no need to design four-, five-, or longer base codons to accomplish the same job performed buy a three-base codon.

Nature makes use of the extra 44 codons that do not translate directly to an amino acid by assigning some of them specific meanings, such as "The gene starts here" and "The gene stops here." The genetic code, unlike the 26-letter alphabet used in English, contains redundancy but no ambiguity. Redundancy allows some amino acids to have more than one codon that defines them. For example, DNA uses either of two codons to spell the amino acid arginine (AGA and AGG), but six different codons can each spell the amino acid serine. No ambiguity occurs in the genetic code, however, because no codon ever specifies more than one amino acid. Contrast the genetic code to the English alphabet containing the five-letter heteronym "spring," which could mean a mechanical device inside a mattress, a freshwater source, the act of leaping, or a season.

Redundancy helps biological systems operate with some versatility so that even a slight mistake in a base sequence can translate into the correct amino acid used for building a protein. Cells also contain repair systems that proofread the code. Repair system enzymes excise incorrect bases, fix mismatched bases in the ladder's rungs, and rebuild damaged sections of DNA.

The genetic code connects all biological organisms. Regardless of the organism from single-celled bacteria to the most complex—usually assumed by egocentric humans to be the human—all use the same genetic alphabet to define amino acids and thus proteins. The universal nature of the genetic code allows scientists to study *E. coli* for the purpose of learning about human genes. In addition, the unity of biology makes the opportunities for genetic engineering almost unlimited because every organism uses the same basic means of building its cellular constituents.

Genetic engineering has not replaced the chemical industry, although industry leaders in Europe have made plans to convert chemical manufacturing processes to biological processes. This new business model, called white biotechnology, uses bacteria or their enzymes to carry out manufacturing steps that presently require high heat and hazardous catalysts. White biotech produces no hazardous waste and requires much less energy input than conventional manufacturing. The U.S. biotech industry familiar to most people and responsible for making GMOs is called green biotechnology. The biotechnology industry currently designates color codes to specific areas of interest:

- **Green**—Bioengineered microbes, food crops, and trees
- **White**—Microbial enzymes applied to industrial manufacturing
- **Blue**—Biotechniques oriented toward marine biology
- **Orange**—Engineered yeasts
- **Red**—Medical gene therapy, tissue therapy, and stem cell applications

In the 1950s, companies rebuilt their businesses for a peacetime economy. The chemical industry had been expanding since the 1930s and boomed in the 1950s with a new mantra: convenience products. DuPont Company communicated its industry's bright future as well as its customers' with the slogan "Better Things for Better Living... Through Chemistry." By 1964's New York World's Fair, DuPont had rolled out a song-and-dance extravaganza on the power of chemistry. The industry's new drugs, pesticides, and plastics promised a better quality of life, but these products also required significant quality control at the manufacturing level. Wartime expertise in physics and chemistry turned toward making new analytical equipment to inspect a compound's structure and measure its purity. Companies such as Hewlett-Packard, Varian Associates, and Perkin-Elmer filled the gap.

Alec Fleming's legacy inspired a new interest in biology in the 1940s, but additional breakthroughs came slower than many people might have expected. Antibiotic discovery involved laborious manual tests. Microbiologists scooped up soil samples, recovered the soil's fungi and bacteria, and then searched for extracts from the cultures to test against hundreds of bacteria. In addition to the tedium, microbiologists often saw variable results in laboratory tests. When a microbiologist inoculates ten tubes of broth with the same *Staphylococcus*, eight tubes might grow, one tube might not grow, and the tenth tube ends up contaminated. Chemists at drug companies sped up the process by synthesizing new antibiotics based on the structures of known natural antibiotics. By the 1950s, the chemical industry offered a faster way to find new drugs. To keep up with the chemists, microbiologists needed a dependable microbe that grew easily and quickly to large quantities.

## E. coli

In the 1880s, outbreaks of infant diarrhea raged through European cities and killed hundreds of babies. Like other physicians, Austrian pediatrician Theodore von Escherich struggled to save his patients and simultaneously find the infection's cause. He recovered various bacteria from stool samples without an idea as to their role, if any, in the illness. In 1885 von Escherich published a medical article describing 19 bacteria that dominated the infants' digestive tracts. One in particular seemed to be consistently present and in high numbers. He named it (with a striking lack of creativity) *Bacterium coli commune* for "common colon bacterium." In 1958, the microbe was renamed *Escherichia coli* in honor of its discoverer.

*E. coli*'s physiology offers nothing remarkable. It does not pump out an excess of useful or unique enzymes or make antibiotics. It dominates the newborn's intestinal tract but gradually other bacteria overtake it and carry out the important microbial reactions of digestion. For example, strict anaerobes produce copious amounts of digestive enzymes that help break down proteins, fats, and carbohydrates. These bacteria also partially digest fibers and synthesize proteins and vitamins that are used in the host's metabolism. *E. coli* does not contribute as much to digestive activities as the strict anaerobes, but because it is a facultative anaerobe that uses oxygen when present and lives without oxygen in anaerobic places, its main role is to deplete oxygen so that anaerobic bacteria can flourish.

Von Escherich probably noticed that *E. coli* grows fast to high numbers in laboratory cultures. The species flourishes on a wide variety of nutrients and does not need incubation. Leave a flask of *E. coli* on a lab bench overnight and a dense culture will greet the microbiologist the next morning. The strict anaerobes of the digestive tract take three days or longer to grow to densities that *E. coli* reaches in about 10 hours.

By the turn of the century, doctors had not solved the problem of infant diarrhea—it remains a significant worldwide cause of infant mortality. They did, however, think that *E. coli* might be useful for treating intestinal ailments in adults. In Freiburg, Germany, physician Alfred Nissle planned to use *E. coli* for intestinal upsets such as diarrhea, abdominal cramping, and nausea. In this so-called "bacteria therapy" Nissle believed that dosing an ill person with live *E. coli* might drive the pathogenic microbes from the gut.

From 1915 to 1917, Nissle tested various mixtures of *E. coli* strains in Petri dishes against typhus-causing *Salmonella*. When a mixture appeared to be antagonistic toward the *Salmonella*, he tried it on other pathogens. Nissle finally concocted a "cocktail" of what he considered the strongest *E. coli* strains and with considerable courage he drank it. When no harmful effects ensued, Nissle felt he was on the road to an important medical discovery.

During the period Nissle had been conducting his *E. coli* experiments, Germany's army suffered from severe dysentery as had others throughout Europe during the Great War. Dirty water, bad food, and exhaustion conspired to weaken men in the foxholes as well as civilians. In 1917 Nissle made his way to two field hospitals in search of a super *E. coli* that would work even better than the strains in his laboratory. In one tent, he found a non-commissioned officer who had suffered various injuries but never fell victim to diarrhea even when everyone around him had it. Nissle cultured some *E. coli* from the soldier and returned to Freiburg.

Alfred Nissle grew the special *E. coli* in flasks and then poured it into gelatin capsules. When the job of supplying an entire army overwhelmed him, he commissioned the production to a company in Danzig. The new antidiarrheal capsules were called Mutaflor. The wartime upheavals in Europe through 1945 forced Nissle to move the manufacturing more than once, but the production of Mutaflor never ceased. Mutaflor remains commercially available today as a probiotic treatment for digestive upset. The product contains "*E. coli* Nissle 1917" made from direct clones of the superbug Nissle isolated on the battlefield in 1917. The original strain that Nissle submitted to the German Collection of Microorganisms remains in its depository in Braunschweig.

At Stanford University in 1922, microbiologists noted another fast-growing *E. coli* strain with a curious trait: It did not cause illness in humans. The strain received a laboratory identification of "K-12." K-12 became a standard in teaching and research laboratories, soon shared by Stanford with other universities. When eventual Nobel Prize awardees Joshua Lederberg and Edward Tatum began studies on how genes carry information and the mechanisms organisms use to exchange this information, they made the logical choice of K-12 as an experimental workhorse. *E. coli* became forever linked with advances in genetics and biotechnology.

Since the first K-12 experiments, more than 3,000 different mutants of this bacterium have been used in cell metabolism, physiology, and gene studies. One of the first bacterial genomes to be sequenced was that of K-12; the complete sequence of its 4,377 genes was published in 1997. In the last 50 years, 14 Nobel Prizes have been awarded based on work done with *E. coli*, mainly K-12.

## The power of cloning

By the 1970s, microbiologists were routinely taking apart *E. coli* to study its reproduction, enzymes, and virulence. The chemical industry had lost some luster with each new discovery of environmental pollution, and biology again looked like the science for the future: clean, quiet, and nonpolluting.

In biotech's infancy, "cloning" became the buzzword for the power of this new technology: The ability to take a single gene and produce millions of identical copies. *E. coli* became a living staging area in which genes were cloned by the following general scheme:

1. Extract DNA from an organism possessing a desirable trait (a gene).
2. Cleave the DNA into many smaller pieces with a specialized enzyme called restriction endonuclease (RE).
3. Extract *E. coli* plasmids and open their circular structure with another RE.
4. Insert the various DNA fragments into the many plasmids.
5. Allow the bacteria to take the plasmid back into their cells.
6. Grow all the bacteria and use screening procedures to identify the cells carrying the desirable gene.
7. Grow large amounts of these gene-carrying cells, that is, cloning.
8. Harvest the product that the gene controls.

In biotech's infancy scientists painstakingly worked out each of the preceding steps (see Figure 5.1). Molecular biologists perfected the art of extracting DNA from cells without breaking the large molecule into pieces. They devised techniques for splicing new segments of one type of DNA into a second DNA molecule, and they developed methods for testing the activity in a new GMO. But the scientists also noticed that their favorite bacterium *E. coli* resisted taking plasmids into their cells, a key step in genetic engineering called transformation. Without an easy way to deliver DNA into *E. coli*, many genetic experiments might become impossible. In 1970, Morton Mandel and Akiko Higa solved this dilemma by showing that calcium increased the permeability of cell membranes to DNA. By soaking *E. coli* in a chilled calcium chloride solution for 24 hours, biologists now make the bacterium 20 to 30 times more receptive to taking up plasmids. Bacteria that take in plasmids from the environment are called competent cells, and biotechnologists now use this simple soaking step to make *E. coli* competent for transformation.

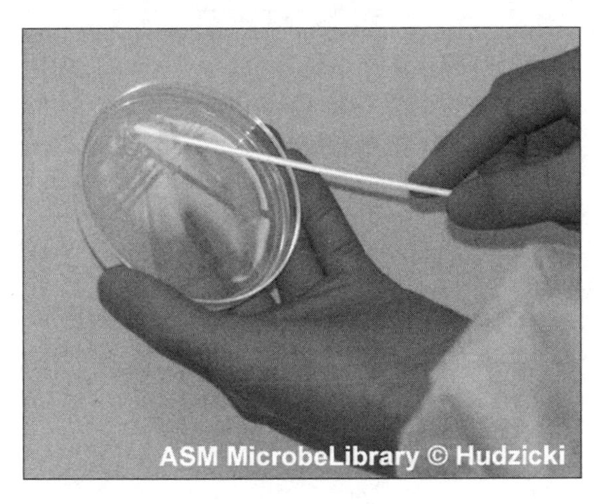


Figure 5.1 Microbiologists in environmental study, medicine, industry, and academia use the same aseptic techniques. These disciplines use methods adapted from biotechnology for manipulating the genetic makeup of bacteria. (Reproduced with permission of the American Society for Microbiology MicrobeLibrary (http://www.microbelibrary.org))

In the early days of biotech research, bacterial cloning—it used to be called gene splicing—served as the only way to make large amounts of gene and gene products. Bacteria make millions of copies of a target gene by replicating it each time a cell splits down the middle to make two new cells, a process called binary fission. In time, scientists developed methods for using bacteriophages to deliver genes directly into a bacterium's DNA. Viruses' *modus operandi* involves appropriating a cell's DNA replication system, which is a perfect mechanism for delivering a foreign gene into bacterial DNA. PCR would enter the picture next as a faster DNA-amplification method.

*E. coli* remains a major tool in biotechnology, but additional microbes such as the yeast *Saccharomyces cerevisiae* and the bacterium *Bacillus subtilis* also contribute a large share to recombinant DNA technology. Biotech companies use the basic cloning scheme described previously in yeasts and bacteria for making the drugs listed in Table 5.1.

**Table 5.1** Major products made by biotechnology

| Made by *E. coli* | Made by Other Microbes |
|---|---|
| Interferons as antiviral and antitumor drugs | Antitrypsin emphysema treatment |
| Colony-stimulating factor to counteract effects of chemotherapy and treat leukemia | Factor VIII for treatment of hemophilia |
| Growth hormone | Bone morphogenic proteins to induce new bone formation |
| Insulin for diabetes | Calcitonin for regulating blood calcium levels |
| Interleukins for treatment of tumors and immune disorders | Erythropoietin for anemia |
| Relaxin as a childbirth aid | Growth factor for wound recovery |
| Somatostatin treatment for acromegaly, a growth disorder of bones | Hepatitis B vaccine |
| Streptokinase as an anticoagulant to reduce blood clots | Macrophage colony-stimulating factor for cancer |
| Taxol ovarian cancer chemotherapy drug | Pulmozyme for breaking down mucous secretions in cystic fibrosis patients |
| Tumor necrosis factor to disintegrate tumor cells | Serum albumin as a blood supplement |

Biotech companies manufacture their products in fermentation vessels of 300 to 3,000 gallons. Technicians at biotech companies scale up bacterial cultures from small volumes of less than a gallon to fermenters of several gallons. After this modest scaling up of the culturing process, workers in a manufacturing plant increase the production size more by growing the GMO in vessels of 300 to 3,000 gallons. All of the actions leading up to large-scale production comprise upstream processing. A different team of technicians monitors downstream processing, which encompasses all the steps from fermentation to the packaging of a clean, pure final product. Genentech and Amgen, both in California, became the first two biotech companies to reach this large-scale level of production.

In 1996, scientists at Scotland's Roslin Institute created Dolly, the first mammal (a sheep) made by cloning DNA from an adult animal. The public and many scientists reacted to this news with concern that humans would be the next organism to be cloned. Renee Reijo Pera, stem cell researcher at the University of California-San Francisco, remarked, "You can almost divide science into two segments: Before Dolly and After Dolly." But cloning higher organisms had little in common with bacterial cloning for making GMOs. Dolly's clone came about by transferring the nucleus—containing an animal's entire genome—from an adult sheep's cell into mammary tissue where the genome replicated as the tissue reproduced. Cows, goats, pigs, rats, mice, cats, dogs, horses, and mules have since been similarly cloned.

The goal of animal cloning is to produce a new animal identical in every way possible to the original animal. Animal cloning thus seeks to repeat an entire genome in a new animal. Gene cloning in bacteria, by contrast, serves as a simple way to make many copies of one or more genes in a short period of time. In short, animal cloning makes new animal copies, and bacterial cloning makes new gene copies. By inserting one or more genes into a bacterial cell's DNA and then growing the cells through several generations, a microbiologist can produce millions of copies of the "new" DNA overnight because of bacteria's fast growth rate.

## A chain reaction

One spring evening in 1983, biochemist Kary Mullis drove from his job at Cetus Corporation near San Francisco to his cabin in California's quiet Anderson Valley. The San Francisco Bay Area had just begun to plant the seeds for the new science called biotechnology. Molecular biologists had learned how to open DNA molecules with enzymes and insert genes from an unrelated organism. But cloning bacteria to make each new batch of genes required considerable labor, and the bacterial cultures produced only miniscule amounts of desired DNA. Mullis pondered this problem as he drove Route 128. He recalled reading of a bacterium living in hot springs and containing enzymes active at high temperatures that melted most other enzymes. Before reaching his cabin, Kary Mullis had developed an idea that would revolutionize biology.

In 1966, famed microbial ecologist Thomas Brock and graduate assistant Hudson Freeze had discovered a bacterium surviving the blistering conditions of Mushroom Spring in Yellowstone National Park. They named the species *Thermus aquaticus*, *Taq* for short, and sent a culture to a national repository for microbes near Washington. DC. Various microbiologists studied the thermophile and its enzymes, but *Taq* seemed to offer little in the way of useful attributes. Mullis suspected that *Taq* in fact held a crucial attribute.

At a temperature approaching 200°F, DNA becomes unstable and separates, or denatures, into two single strands instead of its normal double stranded confirmation. Back in his lab, Mullis raised the temperature of a DNA mixture to denature the molecule and then added DNA fragments called primers plus the enzyme DNA polymerase he had extracted from *Taq*. Next, Mullis lowered the temperature to about 154°F wherein the polymerase began building new DNA copies from the old strands and the primers. By repeatedly heating and cooling the mixture, Mullis could produce about one million copies of the new DNA in 20 minutes and a billion copies in 30 minutes. Molecular biologists call this production of millions of DNA copies from a small, single piece of DNA, amplification. *Taq*'s DNA polymerase provided the key to Mullis's invention because it withstands repeated heating to very high temperatures and then carries out the DNA synthesis step at the cooler (but still high) temperature.

The new process called polymerase chain reaction (PCR) made any snippet of microbial DNA analyzable. Michael Crichton capitalized on PCR's extraordinary potential in the 1990 book *Jurassic Park*, in which scientists amplify dinosaur DNA preserved in ancient amber. Although PCR can amplify pieces of DNA that had been dormant in nature for years, *Jurassic Park*'s re-creation of an entire extinct genome seemed implausible when the movie was released because closing missing gaps was error-prone. Today, computer programs calculate the likely base sequences of missing pieces of DNA to fill in gaps in damaged DNA. As the power of these programs increases, scientists will reconstruct extinct DNA with increasing accuracy.

When a character on a television crime show says, "We need a rush on the DNA," a harried lab technician produces within minutes the name (accompanied by an up-to-date photo) of the bad guy on a computer screen. These scenes depict the power of PCR for analyzing biological matter, but the entire PCR process actually takes much longer. A technician first prepares the DNA-primer-polymerase mixture. A heating-cooling machine called a thermocycler requires at least 2 hours to amplify the bits of DNA. Next, the scientist must determine the sequence of the DNA's subunits or bases, which takes another 24 hours if using an automatic sequencer instrument. Lacking such an machine, manual methods can take up to 3 weeks.

The Food and Drug Administration (FDA) and other government agencies have put PCR to use in crime-solving. In early 2009, the FDA began a food product recall that would encompass 3,900 peanut butter products as suspect causes of a nationwide *Salmonella* outbreak that sickened 700 people and killed nine. Microbiologists used PCR to amplify the DNA from bacteria in the products and determine the pathogen's unique sequence. This so-called DNA fingerprint led CDC investigators to a Blakely, Georgia, manufacturing plant. A leaky roof had allowed rain contaminated with *Salmonella*-laced bird droppings to land directly on food processing equipment and perhaps directly into peanut butter paste as well. Microbiologists can now trace a single pathogen strain from a person's stool sample to an individual farm, a certain shift on the packing line, and even a specific agricultural field.

Real-time PCR has come on the scene as a faster way to analyze samples to prevent crimes from growing cold. In real-time PCR, a detector monitors the formation of increasing amounts of DNA in the thermocycler as it occurs, unlike traditional PCR that takes extra days to analyze the final products from the thermocycler step. Real-time RCR has helped fight the global poaching industry that trades hides, pelts, internal organs, horns, feathers, and shells of endangered animals, as well as caviar. Microscopic drops of an animal's blood on a suspected poacher's clothes can solve a case. Analysis of ivory sold on the black market has traced the ivory to specific elephant herds in Africa and sometimes to individual families.

Kary Mullis received the 1993 Nobel Prize in chemistry for developing PCR technology. Shortly afterward, the enigmatic Mullis joined a campaign to question the idea that HIV causes AIDS.

## Bacteria on the street

Biotech has developed into a science of contradictions. For instance, few graduate students in microbiology complete their studies without doing some sort of gene sequencing or gene engineering. The majority of these students have no exposure at all to culturing whole bacterial cells except *E. coli*, and they spend more time with the disassembled bacterium than the whole living cell. The biotech business has a similar dichotomy. The earliest biotech advocates touted gene cloning as a step toward curing humanity's worst diseases while antibiotech groups predicted the end of nature as we know it. Government leaders recognized the upside for the United States as a world leader in the emerging technology, but they also worried about the need to contain the frightening creatures about to emerge.

A conflicted Wall Street put a modicum of trust in the new industry but did not leap into the biotech pool with both feet. The public's worry over safety did not make for attractive investments. Biotechnology's proteins and cells rather than widgets also presented a new business model. What is a marketable product from a biotech company? Is it the cells that produce a hormone, the gene that encodes for the hormone, or the hormone itself? The U.S. Supreme Court helped clear up some of the confusion by ruling in 1980 that bioengineered bacteria could be patented.

At the start of the 1990s, biotech stocks rode the high-tech wave on Wall Street. By the mid-nineties, however, meager returns turned off the investors, and their interest in biotech cooled. Biotech-produced drugs were not easy to make. The manufacture of genetically altered organisms could produce surprises for even seasoned microbiologists. Mutant cells, contamination, and the capability of bacteria to shift their metabolic pathways slowed the early progress. Biotech's biggest drawback resided in the fact that people could not make up their minds if the new technology was about to save them or kill them.

Warren Buffett described the perfect product, cigarettes, when he said, "It cost a penny to make. Sell it for a dollar. It's addictive." The dotcom industry grew based on this very philosophy. Biotechnology has not come near matching Buffett's three criteria. Like conventional drugs, biotech products require large amounts of research cash and lengthy clinical testing on human subjects. Some drugs, such as new antibiotics, have become too expensive to develop. If people cannot imagine how they would exist without computer technology, they certainly can and do imagine a world without biotech products. In fact, an increasing number of people in the United States prefer a world without GMOs. Do we need GMOs? Bioengineered tomatoes taste fine, but so do organic, non-GMO varieties. Bioengineered bacteria clean up oil spills, but so do bacteria native to the waves and sands slickened with oil.

Biotechnology received a golden opportunity on March 24, 1989, to show the world the value of GMOs released into the environment for a purpose. The *Exxon Valdez* oil tanker hit a reef that day in Alaska's Prince William Sound and spilled an estimated 11 million gallons of crude oil. Whipped by winds, about four million gallons of foamy crude washed ashore, coating 1,300 miles of coastal habitat for marine organisms, terrestrial animals, and birds. Marine bacteria at the spill burst into rapid growth in response to the influx of nutrients; crude oil provides a digestible carbon source for bacteria as opposed to refined oils. The United States did not permit the release of GMOs into the environment in an uncontrolled manner, so microbiologists could not put to work fast-growing bacteria engineered for oil degradation. They turned instead to bioaugmentation to carry out the largest microbe-based pollution cleanup project in history.

The Environmental Protection Agency's John Skinner pointed out shortly after the spill, "Essentially, all the microorganisms needed to degrade the oil are already on the beaches." Microbiologists accelerated the bacteria's growth rate by adding nitrogen and phosphorus to the soils. Like bacteria fed a nutrient-rich broth in a laboratory test tube, the native bacteria responded to the augmentation of their environment with added nutrients. The bioaugmentation of the shore's native bacteria (mainly *Bacillus*) is believed to have increased by at least sixfold the rate of oil decomposition.

GMOs fitted with the genes from native bacteria that degrade fuels, pesticides, industrial solvents, and toxic metal compounds have been developed in microbiology laboratories all over the world. Government agencies have slowed this progress by limiting GMOs to experimental study rather than real-life environmental disasters. The U.S. Environmental Protection Agency reserves the term "bioremediation" for pollution cleanup by unaltered native bacteria and not bioengineered strains. Communities may be nearing a point at which they must choose between hazardous substances in their environment and GMOs released to clean up those substances.

Biotechnology critics have warned of a future in which foods come from 3,000-gallon vats of bacteria. Economic and technology consultant Jeremy Rifkin has cautioned that bacteria will make soil and farms obsolete. I am not sure how this could ever happen. I suspect Rifkin is speaking in metaphor to warn us of GMO-produced foods that could supplant traditional agriculture. Rifkin's Web site warns also that "the mass release of thousands of genetically engineered life forms into the environment [will] cause catastrophic genetic pollution and irreversible damage to the biosphere." Biotechnology today must stand up to strong criticism even as it develops new life-saving drugs and invents processes that clean the environment.

The concept of developing microbial food for humans relates mainly to single-cell protein, or the use of microbial cells as a dietary protein supplement. This idea is at least 20 years old and had been proposed as a way of alleviating global hunger and protein deficiency. Single-cell protein from bacteria never reached practical levels for two reasons. First, bacteria grown in large quantities as a food must be cleared of any toxin or antibiotic they might make, which complicates the production process and raises costs. Second, microbial products packaged as a protein-dense food would likely induce serious allergic reactions in many consumers. Bacteria cannot replace traditional agriculture even if a future generation of scientists found a way to do it. The Earth needs green plants as much as it needs bacteria.

Warnings by biotech's critics about GMOs invading natural ecosystems continue. Bacteria are survivors because of supreme adaptability. Could the adaptability needed by a GMO to carry out its job in nature also allow the microbe to take over ecosystems? Natural ecosystems possess exquisite mechanisms to ensure balance between competitive species from bacteria to higher organisms. Motility, quorum sensing, spore formation, and antibiotic production are examples of the many devices bacteria use to ensure they receive adequate habitat, nutrients, and water. A GMO must overcome all competitors to take over an ecosystem, but nature long ago developed mechanisms to protect the balance of species and resist drastic change. It is wise to remember also that bacteria have been exchanging genes that help them survive since the beginning of their existence. Most new genes that become part of a cell's DNA by either gene transfer or mutation give no benefits to the cell. Because the genes of GMOs are designed to accomplish very specific tasks, the chances of a GMO ruling over natural communities seem remote.

The National Institutes of Health (NIH) have published hundreds of pages of regulations on GMOs with the intent of reducing the chances of a GMO accidentally escaping into the environment. These rules cover methods for containing recombinant microbes by physical, chemical, and biological tactics. Current physical containment involves the safe handling and disposal of GMOs so that live cells do not accidentally escape a laboratory and enter an ecosystem. Microbiologists use special safety cabinets based on the principles of BSL-4 cabinets. They also sterilize all wastes before discarding them. Chemical methods include disinfectants and radiation to kill bacteria in places they might have contaminated. But the dexterity with which bacteria evade chemicals—imagine the consequences of a GMO lodged inside a biofilm—highlights the weaknesses of chemical containment. To date, biological methods for making GMOs safe in the environment offer the greatest promise.

Microbiologists can engineer bacteria to self-destruct by adding suicide genes to recombinant DNA. Suicide genes control GMOs after the microbe has completed its task. The safety mechanism works by either positive or negative control. In both cases, a second compound, or activator, keeps the suicide gene from working until conditions change in the environment. In positive control, a chemical or other stimulus such as a certain temperature, affects the activator, which then releases its control over the suicide gene. The now active suicide gene initiates a progression of events in the cell that lead to its death, a process called apoptosis. In the hypothetical example already mentioned, a 3,000-gallon batch of bioengineered *E. coli* making growth hormone, ruptures and spills in a 7.0 Richter earthquake—a believable event in California where hundreds of biotech companies exist. The *E. coli* might rush into nearby soils and streams, producing hormone that traumatizes ecosystems. But a suicide gene designed to turn on when cells are exposed to a temperature of 72°F or lower—fermenters usually run at about 100°F—ensures that the *E. coli* destroys itself as soon as it escapes into the environment. Negative control turns on when a stimulus in the environment disappears. Bioremediation bacteria designed to degrade a pollutant, for example, undergo apoptosis when the pollutant disappears.

*E. coli* is the world's most bioengineered microbe and also provides suicide genes for other GMOs. The *gef* gene in *E. coli* encodes for a 50-amino acid protein, small by protein standards, that turns on apoptosis in several different bacterial species. The *gef* gene has already been investigated as a treatment against melanoma cells and breast cancer, and for controlling engineered *Pseudomonas*. Bioengineered *P. putida* degrades alkyl benzoates, which are thickeners used in cosmetic products and drugs. As long as this pollutant remains in the environment, *P. putida* equipped with *E. coli*'s *gef* gene continues breaking it down. When the pollutant level has been reduced, the *gef* protein interferes with the normal flow of energy-producing electrons in *P. putida*'s membrane. The bioengineered *Pseudomonas* commits suicide.

Biological containment systems control GMO cells from within the cell and thus promise the best method of preventing GMO accidents. But one *P. putida* cell per every 100,000 to 1,000,000 per generation is a mutant that resists the *gef* gene's action. Will the clones win or will the mutants win? Biotechnologists have helped push the odds in favor of "good" clones over "bad" mutants by inserting two *gef* genes into *P. putida*, which lengthens the odds of resistance to one cell in every 100,000,000.

## Anthrax

If *E. coli* is the world's most bioengineered bacterium, *Bacillus anthracis* is the most feared because it causes the disease anthrax. *B. anthracis* joins various viruses, parasites, other bacteria, and toxins (made by bacteria or fungi) on a list of potential bioterrorism threats. Not only does the *B. anthracis* toxin cause lethal effects in humans, but the bacterium's ability to form endospores helps it out-survive other pathogens. Endospore-formation keeps the cells alive and yet resistant to chemicals, irradiation, and antibiotics.

*B. anthracis* begins as a laboratory culture like other bacteria. To make endospores, a microbiologist stresses the cells by heating the culture broth. The cells begin forming endospores within minutes of this stress. The microbiologist can freeze-dry the spores to make a brown to off-white powder; the color depends on the type of medium that had been used for growing the cells.

The dry, odorless, and lethal powder has caused significant concern in the United States, especially since anthrax was used as a presumed weapon distributed in mail in 2001. Security teams at airports and public buildings now search for unidentified powders as possible anthrax.

As a bioweapon, other pathogens work better than anthrax. The pathogen causes illness if it enters the body through a wound in the skin, by ingestion, or inhalation, with inhalation being the likely route of infection for a bioweapon. The skin route would be impractical for a terrorist and putting anthrax into food or water becomes ineffective because of a phenomenon called the dilution effect. Community water supplies and food products contain such large volumes that a terrorist would find it impossible to contaminate either with a dose big enough to kill. The endospores require a large dose to cause infection in people, so food and especially water would dilute them to harmless levels. A terrorist would furthermore be hard-pressed to perform the laborious culturing and freeze-drying steps needed to make a significant amount of endospores.

Disease by inhalation has caused greater concern because it has already been shown to cause most anthrax cases, the contamination of postal letters in 2001, for instance. But not everyone who gets infected develops disease. People who do get sick cannot transmit it to others because anthrax is noncontagious. Even though *B. anthracis* grows easily in a laboratory, all other characteristics of this microbe make it a poor bioweapon. Therefore, the most feared bacterium is not as big a threat to a large population of people as many believe.

## Why we will always need bacteria

White biotechnology offers the greatest hope of integrating bacteria into industry in a way to positively affect the environment. The use of bacteria to perform activities now carried out by strong acids and organic solvents will drastically cut down on the chemical wastes flowing into rivers, soils, and groundwaters. Many industrial steps must take place at several hundred degrees, which consumes large amounts of energy. Bacteria substitute biodegradable enzymes for caustic chemicals and work at mild temperatures, and they do it quietly. Microbial fermentations also produce heat that can be rerouted into the manufacturing facility to reduce overall energy use.

Bacteria are white biotech's raw material. Rather than watch truck or trainloads of chemicals roll toward manufacturing plants, neighbors of a white biotech company might spot a person carrying a single vial of freeze-dried bacteria. From that point, the bacteria regenerate themselves. In fact, ancient societies might wonder why present-day industry bothers with their noxious mix of materials and wastes. Bacteria already make almost every compound humans find important, even plastic. *Pseudomonas* species make polyesters called polyhydroxyalkanoates (PHAs) from sugars or lipids found in nature. The bacteria use the large compounds as a storage form of carbon and energy and as the sticky binder in biofilm.

Industrial interest in PHAs increases or decreases with oil prices because oil serves as a cheap precursor for making most plastics. As oil prices rise, PHAs become more attractive for making soft containers such as shampoo bottles. But PHA production is not inexpensive due to the costs of nutrients for growing the bacteria and methods for harvesting the polymer.

*Bacillus megaterium* and *Alcaligenes eutrophus* lead a group of diverse species that produce nature's most abundant PHA, polyhydroxybutyrate (PHB). Bacteria excrete higher amounts of PHB under stress, probably as a protective coating around the cells. The narrow environmental conditions that induce the bacteria to turn on their PHB genes make this a very expensive natural product compared with plastics derived from fossil fuels. PHBs are compatible with human tissue because they do not cause allergic reactions, and they are pliable. These attributes make PHBs good choices for medical equipment such as flexible tubing and intravenous bags. To reach this promising future for biodegradable plastics, white biotechnology will be called upon to find the secrets of bacterial metabolism that lead to the cost-effective production of PHAs.

Some manufacturing processes have changed little since the dawn of the Industrial Revolution. Of all aspects of society, manufacturing lags the furthest behind in converting traditional processes into more sustainable methods. For this important change to take place the most self-sufficient organisms on Earth might well lead the way.

# 6

## The invisible universe

The field of microbial ecology focuses on the role microorganisms play in all of nature. Microbial ecologists study bacteria in small habitats of a few dozen species as well as global systems that circulate elements through continents and oceans. These systems called biogeochemical or nutrient cycles make carbon, nitrogen, sulfur, phosphorus, and metals available for humans and all other life. Microbial ecology now includes technologies aimed at reversing global warming, pollution, and biodiversity loss.

Microbial ecologists continually uncover new Earth-human-bacteria relationships. Despite the importance of good bacteria in the environment, microbial ecology is a new science compared with food and medical microbiology.

During the Golden Age of Microbiology, battles against disease inspired microbiologists more than finding out what grew in a clump of dirt. Joseph Lister introduced aseptic techniques for surgical procedures, Edward Jenner developed the smallpox vaccine, and Florence Nightingale promoted hygiene practices for preventing infection. It may have seemed as if the only good bacterium was a dead bacterium.

Late in the Golden Age, botanists Martinus Beijerinck and Sergei Winogradsky took roads less traveled by studying the beneficial bacteria of soil and water. In the Netherlands Beijerinck studied the symbiotic relationships between plants and bacteria. Winogradsky, from Russia, explored bacterial metabolism in soil and water.

Martinus Beijerinck was born in 1851 and grew up in modest surroundings as the son of a tobacco farmer. After pursuing an education in botany and agriculture, he became head of the Netherlands' first laboratory devoted to industrial microbiology. At this position Beijerinck investigated contagious viruses that infected tobacco plants and nitrogen-metabolizing bacteria that live in association with legume plants.

In 1888, Beijerinck discovered bacteria living inside small lumps or nodules on the roots of *Vicia* and *Lathyrus* (yellow pea) plants. Beijerinck performed the difficult tasks of isolating these bacteria from the nodules and growing them in his laboratory. He took on the painstaking process of formulating a nutrient mixture to favor the root nodule bacteria while inhibiting thousands of other bacteria in soil. This method, called enrichment medium, remains a key part of environmental microbiology. Beijerinck spent several years piecing together the metabolism of these bacteria (later to be named to the genus *Rhizobium*) and their role in nature.

Martinus Beijerinck revealed what is now known to be a critical step in the Earth's nitrogen cycle: *Rhizobium* pulls nitrogen from the air, a process called nitrogen fixation, and converts the element into a form that legume plants (peas, beans, peanuts, and alfalfa) can use. The plant incorporates the nitrogen into proteins, nucleic acids, and vitamins, which a diversity of animal life then takes in for nutrition. The *Rhizobium*-legume union represents symbiosis in which two unrelated organisms live in close association. In this case, the type of symbiosis is termed mutualism because both organisms cooperate in giving each a benefit. The root gives the bacteria a safe haven, and *Rhizobium* supplies the plant with an essential nutrient. Not all types of symbiosis are as beneficial as mutualism:

- **Commensalism**—One organism benefits and the other receives neither a benefit nor harm.
- **Amensalism**—One organism benefits by exerting a harmful effect on another.
- **Parasitism**—One organism living on or in a host organism benefits at the expense of the second organism's health.

Beijerinck also studied the sulfur cycle in soil bacteria. The step called sulfate reduction occurs in anaerobic places in soil. Beijerinck devised methods for growing fastidious sulfate-reducing bacteria, a feat that other microbiologists had believed to be too difficult or even impossible.

Winogradsky, born 5 years after Beijerinck, enjoyed a more privileged upbringing. Young Sergei found classes in Greek and Latin "not only uninteresting and unpleasant, but depressing, both physically and mentally." As he grew older he tried law, then music, but inspired by neither he turned to the natural sciences. In 1885, Winogradsky took a position in botany at the University of Strasbourg and began at once to study the sulfur-using bacterium *Beggiatoa*, the bacterium that shuttles between sunlit and dark layers in microbial mats.

Louis Pasteur offered Winogradsky a position at his famed research institute in Paris, but the Russian declined, preferring to return to his homeland to build the microbiology profession there. The Great War disrupted the progress of most professions; in 1917, wealthy families such as Winogradsky's barely escaped death at the hands of the Bolsheviks.

Winogradsky took a position at the University of Belgrade where no science laboratories or even a library existed, but at least it provided some stability for his family. He perused the only scientific journal he could find, *Centralblatt dür Bakteriolge* and thus kept abreast with bacteriology research in Europe. Few microbiologists were examining in depth the bacteria of natural environments. He plunged into studies on the organism he knew best, *Beggiatoa*, examining the microbe's use of iron compounds for energy. The Institut Pasteur called again, and this time Winogradsky accepted, perhaps tempted by the well-funded and stocked laboratories in Paris.

During his career, Winogradsky would discover at least eight new bacterial species in addition to *Beggiatoa*: endospore-forming *Clostridium pasterianum*; the gliding, cellulose-digesting *Cytophaga* of freshwater, estuarine, and marine habitats; and nitrogen-metabolizing *Nitrosococcus*, *Nitrosocystis*, *Nitrosomonas*, *Nitrosospira*, and *Nitrobacter*. The five nitrogen-utilizing bacteria differed from those studied by Beijerinck: these bacteria live free in soil and run separate steps in the nitrogen cycle from those carried out by *Rhizobium*.

Like Beijerinck, Winogradsky studied the sulfur bacteria and became the first microbiologist to isolate pure cultures of sulfur-oxidizing bacteria from soil. These bacteria turn the element sulfur into a usable inorganic form that Beijerinck's bacteria then convert to a molecule useful to higher organisms. A bit of a Renaissance man of microbiology, Winogradsky also became the first bacteriologist to study biofilms in aqueous habitats, and he sparked interest among microbiologists in iron-metabolizing bacteria living in deep aqueous sediments.

Winogradsky continued writing on microbial ecology into his nineties. His daughter Helen would join him at the institute and carry on his work on nitrogen-using bacteria after his death at age 97.

## Versatility begets diversity

Communities such as biofilms and microbial mats make life easier for their members than living alone as a single cell. But all species of bacteria spend some part of their existence free from a microbial community. Cells break away from communities when the density grows too high. Motile cells escape toxins by themselves or migrate toward nutrients by using flagella, cilia, or twitching movements. During the periods of growth in which cells fend for themselves separate from a microbial community, they often meet their toughest challenges for survival.

Bacteria grown in laboratories encounter few of the discomforts found in nature. Rich nutrient broths, incubators set at perfect temperature, and culture vessels bathed in the bacteria's preferred gas make laboratory life plush compared with life in soil or water. In the lab, bacteria grow faster and bigger than in nature.

Out in the real world bacteria confront scant nutrients, inadequate adherence sites, toxic chemicals, and predators. But with diversity comes versatility, and bacteria have developed a multitude of tactics to ensure their survival in the environment.

In nature, bacteria wage constant competition with protozoa, algae, plants, insects, and worms for nutrients in the soil or natural waters. Unlike these eukaryotes, bacteria become dormant, construct an endospore, or select an alternative metabolism to ride out tough conditions. When nutrients are few, bacteria hold cell size to a minimum; cells that in the lab grow to three or four µm in diameter might reach only one to two µm in nature. This downsizing reduces the amount of nutrients a cell needs, increases the number of safe hiding places on surfaces, and might help bacteria to go airborne and thus move to better environments. Small size also leads to faster reproduction so that a species survives in part by producing enormous numbers of progeny.

Single bacterial cells weather the harsh conditions in their environment and rejoin a community as soon as they can. Part of community-building involves the ability to stick to surfaces. Pathogens and nonpathogenic bacteria both rely on adherence as a key part of their survival mechanism. Like pathogens, environmental bacteria use tiny appendages called fimbrae to attach to things such as rock, soil particles, leaves, or decomposing matter. On surfaces lacking a topography good for attachment, bacteria use electrical charges to help them stick.

Bacteria have a small negative charge on their outside due to the chemistry of the carbon and phosphorus in their proteins and acidic portions of the cell wall. In aqueous environments where most bacteria live, the negative cell attracts positively charged molecules. A negatively charged cell therefore travels through the environment wrapped in a positively charged suit. The minerals in rock and soil also have a positive charge. Organic matter in nature carries a negative charge like bacteria and also attracts its own suit of positive particles. Bacteria would seem to have no chance of adhering to a surface because of all the positive-positive repulsion. But matter behaves differently at the nanoscale level than it does at visible or microscopic sizes measured in micrometers.

At certain nanometer (nm) distances, positive-positive repulsion prevents bacteria from sticking to positively charged objects. At about 10 nm from a surface, a pebble for instance, bacteria detect a small electrical attraction to the surface, but repulsion increases as a cell comes nearer to the pebble. The amount of repulsion wavers due to additional chemical forces existent between 10 nm and 2 nm from the surface. If the cell manages to reach within 1 nm of the pebble, the attractive forces win out and the cell can adhere.

Not only must bacteria overcome the competing chemical forces that occur between 10 nm and 2 nm, they also must find a site not already occupied by other cells, settle in a spot far from microbes secreting antibiotics, and locate a place that affords nutrients, light, and air.

Evading the action of natural antibiotics takes additional guile. Most of the natural antibiotics in use today came from soil microbes. Soil bacteria must resist not only the naturally produced antibiotics in their environment but also synthetic antibiotics that contaminate water coming from human-populated places. Chlorine-containing pollutants, toxic metals (such as mercury, cadmium, silver, and copper), and radioactive chemicals also harm bacteria except for the relatively small number of species that have adapted to these substances. As the amount of pollutants increases in soil, the number and diversity of bacterial species decrease. Adaptations, in fact, provide bacteria with their most powerful survival mechanism. Because of their fast reproductive rate, bacteria can make vital adaptations such as antibiotic resistance part of their genetic makeup more efficiently than any other organism.

After overcoming the travails of starvation, lack of sites to live, and toxic substances, bacteria still must deal with predation. Protozoa roam aqueous environments in nature as they do inside the rumen, engulfing and digesting bacteria. A protozoal cell gobbles 1,000 to 10,000 bacteria for each cell division. Bacteria's greatest defense against extinction is a reproductive rate faster than that of protozoa. The size diversity of bacteria helps, too. Larger protozoa (100 to 1,000 μm in length) capture larger bacteria, leaving most of the small bacteria for small protozoa (5 to 100 μm in length). In many other areas of nature, organisms of similar type diversify the prey they target. Wolves target elk and leave smaller prey such as jackrabbits to the coyotes. This hierarchy of prey and predators ensures the survival of biodiversity. In the microbial world, the protozoa size-to-prey size ratio is about ten to one. On rare occasions, however, protozoa try to take in food larger than their size with deadly consequences.

Certain bacteria in nature prey on other bacteria each in their unique way. *Bdellovibrio* lives in a broad range of habitats from soil to fresh and salt waters and in sewage. This gram-negative genus preys on other gram-negatives by attaching to a cell and secreting enzymes that bore a hole in the cell wall. The predator then squeezes into the space between the prey's cell wall and membrane. The prey cell dies but the *Bdellovibrio* stays and wears it like a coat that somehow resists any new predators.

The aptly named *Vampirococcus* attaches to its prey but does not penetrate the bacterium. It excretes enzymes to partially degrade the prey cell, preferably photosynthetic species. After sucking out all of the prey's cytoplasm, *Vampirococcus* leaves behind an empty cell wall.

Myxobacteria have their own distinctive type of predation. Motile myxobacteria form "wolf packs" of a few dozen to hundreds of cells that glide through the soil in search of prey. The long, thin rods line up in parallel with a few leader cells extending a bit in front of the pack. Myxobacteria packs gracefully patrol the waters for food. After devouring all bacteria in an area, the myxobacteria cells aggregate into a huge funguslike structure called a fruiting body that grows up to 75 millimeters in height. This body, like nothing else in the bacterial world, contains pigments that color the colonies red, orange, yellow, or brown. The fruiting body's stalk raises a sac of cells above the soil's surface. Wind or rain liberates the myxobacteria and carries them to a new location. If conditions at the new site look good, the myxobacteria begin a new life cycle. Fruiting bodies are easy to spot on decaying organic matter, particularly beech and elder trees.

Microbial ecologists have not determined the role of predation in the microbial world. Predation certainly benefits predators in places with low nutrient supply. The predator lets the prey do the work of absorbing and concentrating nutrients, and then gulps the entire meal. Some predators take in bacteria but do not digest them. In the termite gut for instance, bacteria inside protozoa inside the insect digest the woody fibers that termites ingest.

Three hallmarks of bacteria contribute to their versatility. First, the huge size of bacterial populations increases the chance of developing mutants with one or more new, favorable traits. Second, short generation times help a species make the new trait part of its genetic makeup. Third, because bacteria are compact, they have developed enzymes that can do more than one function. For example, enzymes that degrade common organic compounds in nature might also decompose pollutants. The principle behind bioremediation is to use microbes that prefer decomposing a pollutant even when other foods are available.

Large numbers of organisms with different nutrient needs, energy generation, and adaptations would be expected to create a diverse population such as found among microbes. Microbial diversity dwarfs that of other life forms, and microbial ecologists suspect that the diversity is highest in the belt that circles the globe near and at the equator. Biodiversity of higher organisms, plant and animal, in the equatorial tropics exceeds that in other regions on Earth. In this region, an abundance of sunlight might lead to higher numbers of photosynthetic bacteria, which give rise to food chains on land and in water. Because of the tropics' higher overall biodiversity, the region is environmentally stable. This allows countless small and specialized populations to exist. Diverse populations in turn offer bacteria more options for creating symbiotic relationships. Finally, a stable tropical climate compared with the seasonal changes of temperate regions gives bacteria better opportunities to evolve and develop useful adaptations.

Microbial ecology challenges microbiologists because the bacteria studied in laboratories are not necessarily the most abundant. This is due to VBNC, meaning viable but not culturable. Craig Venter's genetic analysis of marine microbes supported the idea long suspected by biologists, that microbial diversity is far greater than even the highest estimates. VBNC bacteria either do not grow in lab conditions or microbiologists have yet to discover the things these species need. As a result, microbiology must base most of its theories on how a small minority of the world's bacteria behave in a laboratory. Genetic testing, such as Venter's, will help solve this problem because whole bacteria need no longer be the focus of experimental study. By analyzing gene diversity, microbiologists will learn more about microbial diversity.

## Cyanobacteria

No single bacterium can be thought of as more important than any other, but if pressed to select one above others, I would pick cyanobacteria. These microbes that biologists originally misidentified as blue-green algae almost single-handedly symbolize bacterial diversity.

The bacteria that began providing the Earth with oxygen three and a half billion years ago show wonderful versatility that spans terrestrial and aquatic environments, freshwater and marine. Cyanobacteria (see Figure 6.1) have few constraints on where they live other than needing sunlight for photosynthesis. On land, cyanobacteria often pair with fungi to form lichens that live on inorganic surfaces. The cyanobacterium *Anabaena* forms a similar relationship in water with *Azolla*, a small floating fern. In this association, the plant supplies about 90 percent of the photosynthesis, and *Anabaena* takes responsibility for pulling nitrogen from the air to supply itself and *Azolla*.

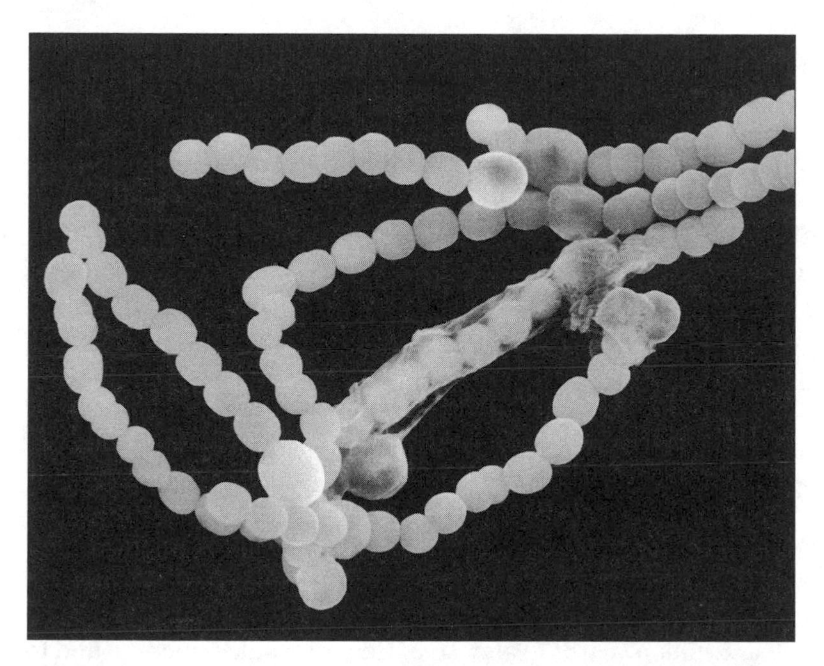

Figure 6.1 Cyanobacteria. Cyanobacteria contain a diversity of species and activities, all having photosynthesis in common. This string of Anabaena cells contains larger cystlike cells that fix nitrogen. (Courtesy of Dennis Kunkel Microscopy, Inc.)

Cyanobacteria dominate microbial mats and attach to terrestrial surfaces (the periphyton form). They live in the greatest abundance, however, as free cells in aquatic habitats and make up a large portion of marine plankton as shown on Table 6.1. At normal ocean concentrations of about 100,000 cells per milliliter, the microbes are invisible, but in larger, denser populations called blooms, cyanobacteria can turn the waters red, perhaps the inspiration for naming the Red Sea. The oceans receive ample sunlight at the surface, but the light penetrates no deeper than about 320 feet. For this reason, cyanobacteria and all of the world's marine photosynthesis occurs in this layer called the photic zone.

**Table 6.1** Main constituents of marine plankton

| Organism | Cells per milliliter of seawater |
|---|---|
| Krill | Less than one |
| Algae | 3,000 |
| Protozoa | 4,000 |
| Photosynthetic bacteria | 100,000 |
| Heterotrophic bacteria | 1,000,000 |
| Viruses | 10,000,000 |

Green plants, algae, and cyanobacteria act as the Earth's main conduits for converting the sun's energy into usable energy for animals. The oceans contribute the major share of this process. Just as cyanobacteria provide the energy that powers the metabolism of microbial mats, they play a similar vital role as the foundation of marine food chains. From the marine or freshwater cyanobacteria, energy transfers to small organisms and to progressively larger animals until the food chain reaches the top predator, referred to as the "top of the food chain."

As Earth's oldest bacteria still in existence, cyanobacteria played a part in the rise of algae, primitive plants, and today's higher plants. The oldest known fossils are of cyanobacteria from the Archaean period before oxygen began accumulating in the atmosphere. These fossils dated to 3.5 billion years are almost as old as the oldest rocks, dated to 3.8 billion years.

During the Archaean and Proterozoic Eras, cyanobacterial photosynthesis changed the atmosphere's composition from oxygenless to oxygenated. Sometime during the transition from the Proterozoic to the Cambrian Era, some of the large cells dependent on oxygen engulfed a few cyanobacterial cells. A portion of the engulfed cyanobacteria managed to resist being digested inside the predator, and they evolved with succeeding predator generations to become an organelle of the host cell. Plant life descending from these primitive cells evolved into more complex structures, and the vestiges of ancient cyanobacteria would become chloroplasts, the sites in plant cells that convert sunlight energy to chemical energy in the form of sugars.

Cyanobacteria grow slower than most other microbes, doubling about once per day, but they nevertheless compete well against other bacteria because of their durability and the capacity to exist on almost no nutrients; cyanobacteria need only sunlight for energy, carbon dioxide for carbon, and miniscule amounts of salts. These bacteria have larger than normal cell size that reaches several µm in diameter, and they possess a more complex internal structure than other bacteria. The cytoplasm contains a network of membranes that support the enzymes and pigments that run photosynthesis. Cyanobacteria cell shapes also present a unique collection of blocks, chains, and long filaments that microscopically resemble algae more than bacteria.

In the 16th century, Swiss physician Paracelsus—his birth name was Aureolus Phillipus Theostratus Bombastus von Hohenheim!—made one of the earliest observations on cyanobacteria. He noted the mucuslike colonies growing on plants and named the growth *Nostoc*, a term generally meaning nasal discharge. If *Nostoc* was the first cyanobacterium studied, the marine species *Prochlorococcus marinus* is one of the newest. Discovered in 1986, *P. marinus* is possibly the most abundant organism of any type on the planet. *P. marinus* is also the smallest cyanobacterium and one of the smallest known bacteria at 0.6 µm in diameter. The species acts as a photosynthesis machine with few other ecologically important activities. It contains only 1,716 genes. Because the ocean conditions change slowly, *P. marinus* can survive with only a few genes to help it respond to its environment.

The best places to see cyanobacteria in nature are rocky shorelines and on seashells. Microbial ecologist Betsey Dexter Dyer described cyanobacteria on shorelines as a slippery, brown-black, velvety coating on rocks. In aquatic environments, the microbe is evident by its blue-green, green, yellow-red, orange, or violet pigments.

Cyanobacteria in all forms serve the Earth as a tremendous storage site for carbon and nitrogen. Photosynthesis converts carbon dioxide to sugar, which the plant uses to make structural fibers and starches. All bacteria that decompose organic matter on land or in the sea add to the large stores of the Earth's carbon.

## Bacterial protein factories

Ruminant animals and to a lesser extent humans and other single-stomach animals (monogastric animals) get a large portion of their amino acid requirements from bacteria. Enzymes in the intestines digest bacteria, and then specific enzymes called proteases break down the bacterial proteins to liberate individual amino acids, which the animal absorbs. These amino acids or the nitrogen they contain serve as the basis for the animal's own protein synthesis.

This process, complicated though it may seem, is but one step in a global nitrogen recycling system. The nitrogen cycle is perhaps the most-studied nutrient cycle due to its importance to agriculture and the health of humans and other animals. Most people do not suffer from a lack of carbon in their diet. Nitrogen in the form of protein is another story. Nitrogen often occurs in limited supply in diverse environments, making the nitrogen cycle all the more crucial for Earth's organisms.

The proteins in a steak dinner result from a global cycle of nitrogen use and reuse. Without exaggeration, the steak's nitrogen may have come from cyanobacteria in a distant ocean or soils on another continent. Nitrogen gas makes up 78 percent of the atmosphere, almost four times as much as the next most abundant constituent, oxygen. Despite this apparent abundance of nitrogen, living things expend much more energy to get nitrogen into their bodies than they do to absorb oxygen. Except for bacteria, no life takes nitrogen gas directly into the body like oxygen enters. Bacteria, including cyanobacteria, make nitrogen available for all other life by taking in the gas in a process called nitrogen fixation and converting the nitrogen into a form usable by plants. Grazers such as rabbits convert the plant nitrogen (mainly in vitamins and nucleic acids such as DNA) to animal nitrogen (mainly as protein in muscle).

Some species, such as *Azotobacter* and *Beijerinckia*, live independently in the soil and perform nitrogen fixation there. These bacteria convert the gas to ammonia by adding hydrogen atoms to each nitrogen atom. The bacteria *Rhizobium* (discovered by Beijerinck) and *Bradyrhizobium* also absorb nitrogen from the air, but they do so from inside bumps on the roots of plant cells, called root nodules. Martinus Beijerinck discovered this bacteria-plant process in 1888. The bacteria-plant relationship in nitrogen fixation represents symbiosis, which is the cooperative association of two organisms living in close proximity. The intestinal bacteria in humans, other animals, and insects also illustrate a symbiotic relationship.

After nitrogen gas has been converted into ammonia, the nitrogen passes through a sequence of reactions each carried out by the bacteria that Sergei Winogradsky discovered more than a century ago. *Nitrosomonas* converts the ammonia to compounds called nitrites (two oxygens attached to a nitrogen), *Nitrobacter* turns nitrites into nitrates (three oxygens attached to nitrogen), and plant roots then absorb the nitrates for their own nitrogen needs. An entirely different group of bacteria takes excess nitrates from the soil, turns it into the gas nitrous oxide, and releases this gas back into the atmosphere.

The nitrogen that ends up in plants is used by the plant to build vitamins, nucleic acids, and proteins. Cattle grazing on grasses and clovers take in the plant nitrogen, and then rumen bacteria begin their task of building microbial proteins. Humans benefit from the entire process when they ingest proteins in beef. The environment also receives a share of nitrogen when plants die and decay (by the action of soil bacteria such as Bacillus), releasing nitrogen into the soil, and manure from cattle farms leaches into the soil and surface waters.

The nitrogen cycle, so essential for all life, takes up a lot of space. Meat-producing cattle and sheep take up thousands of square miles of land across the world. Countries with a large land area like the United States or Canada can manage this problem, but water-stressed and tropical regions find that meat production makes impossible demands on their environment. Meat animals compete with humans for water in an increasingly large portion of the world. In the tropics, meanwhile, farmers are cutting down or burning jungle to clear land for cattle. As the tropics shrink so does biodiversity.

Many environmentalists feel that large animal meat production threatens the environment, prompting scientists to investigate bacteria as a direct protein source. The nitrogen cycle will continue running, of course, but humans might put less demand on it by using alternate protein sources. The cyanobacterium *Spirulina* (see Figure 6.2) has drawn interest as a potential microbial protein source. Dried *Spirulina* powder serves as a vitamin and protein supplement. *Spirulina* cells are up to 70 percent protein. Most other bacteria consist of about 50 percent protein. Furthermore, *Spirulina* protein is high-quality protein, meaning it contains all of the essential amino acids for humans. *Spirulina* resembles all other photosynthetic organisms by supplying a variety of vitamins and minerals. The enzymes that run photosynthesis require a constant input of vitamins and minerals that act as co-factors, supplementary molecules that participate in chemical reactions. Consider the following attributes of *Spirulina* as a food:

- More beta-carotene, which the body converts to vitamin A, than carrots
- 28 times more iron than beef liver
- Higher concentration of vitamin $B_{12}$ than any other food.

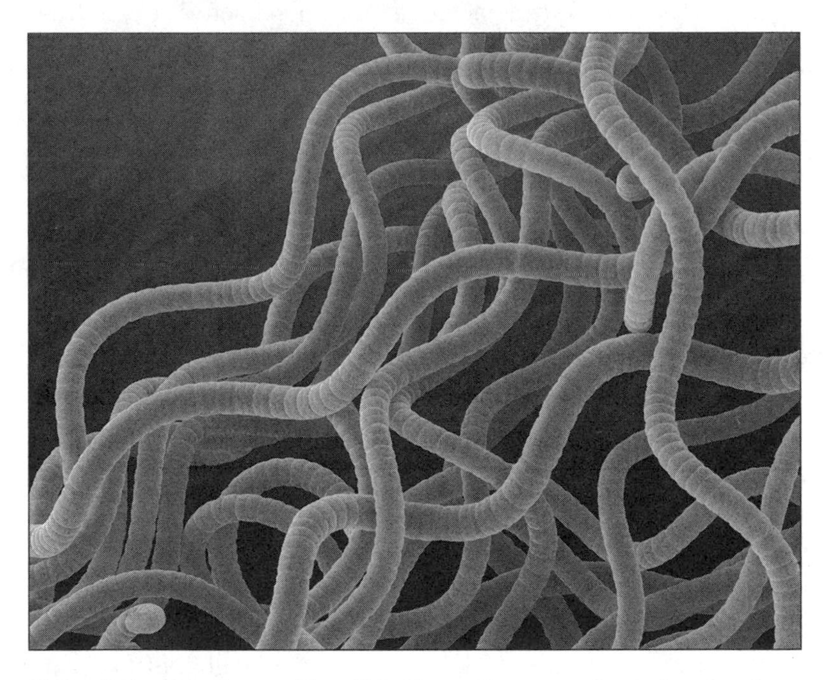

Figure 6.2 *Spirulina pacifica*. This filamentous cyanobacterium has been used for centuries as food. Masses of growth are collected from water, sun-dried, and patted into flat cakes to cook or eat directly. (Courtesy of Dennis Kunkel Microscopy, Inc.)

Does *Spirulina* have a future as a new protein source for undernourished regions of the world? On a per-acre basis, the cyanobacterium supplies 200 times the protein yield of beef and consumes 315 times less water. NASA has experimented with the use of *Spirulina*, misidentified as algae, as a food for space flights. *Spirulina* farms have grown in number worldwide from Thailand, India, and to the United States. These farms contain large ponds for growing the bacteria in a continuous flow system of incoming nutrients and outgoing final product. Microbiologists maintain a narrow range of growth conditions to enhance cyanobacteria growth and inhibit the growth of contaminants.

The environment's current precarious condition requires people to make hard choices regarding the materials they consume. *Spirulina* may become an important aspect of sustainability, but it has not yet arrived there.

## How to build an ecosystem

A pond, meadow, or tidal pool is an example of an ecosystem. Each ecosystem contains a network of interactions between multicellular plants and animals, tiny invertebrates, microbes, and inanimate objects such as soil, water, rocks, and air. Bacteria operate in ecosystems by interacting with other microbes as well as with the immediate microscopic environment composed of liquid and solid surfaces.

In liquids, bacteria contend with the positively or negatively charged substances dissolved or suspended in the microenvironment. Some cells turn on their chemotaxis mechanism to swim toward favorable conditions or away from harmful conditions. Other bacteria float in the milieu and absorb any nutrients they meet, or these cells are swept toward communities of mixed species and settle down there.

In liquid environments with high content of organic matter, bacteria aggregate into a thin film that covers the surface. In this position, cells take in oxygen from the air and absorb nutrients from below. Bacteria in surface films must regulate the surface tension of the air-water interface to stabilize the film's structure and reduce breakup.

Some bacteria, such as *Pseudomonas*, secrete surfactants to help regulate surface tension. These detergentlike substances allow hydrophobic compounds that land on the film to become miscible with the water, and this provides the film with a potential new nutrient source. Surfactants also help nutrients enter microbial cells, thus helping the cells stay in the fragile surface film. Surfactants play a similar role for bacteria in soil on root surfaces.

Soil bacteria live in a microenvironment made mainly of silicon, the Earth's most abundant element. Soil also has substantial amounts of aluminum, iron, calcium, sodium, potassium, and magnesium. Most of these elements occur in a positively charged form that influences the cells' attachment to soil particles. Sometimes bacteria live in a community attached to the inanimate particles, but in other instances they inhabit moist micropores varying from micrometers to millimeters in size. These tiny microenvironments are often the places where nutrients begin cycling through the Earth, atmosphere, oceans, and to every living thing.

The sulfur cycle consists of more chemical conversions than any other known nutrient cycle. Sulfur as sulfur dioxide gas enters the atmosphere when released from volcanic activity, including hot sulfur springs. Fossil fuel combustion also adds large amounts to the atmosphere. Most of the Earth's sulfur is held in the planet's core with lesser amounts in biological matter. The Earth's crust contains almost $2 \times 10^{16}$ tons of sulfur; the terrestrial and marine biological matter holds about $1 \times 10^{10}$ tons.

Two groups of bacteria nicknamed the green sulfurs and purple sulfurs for the type of pigments they contain convert elemental sulfur to sulfate compounds. Elemental sulfur—this is pure sulfur unattached to any other element—is a solid that sticks to soil particles as well as the surface of bacterial cells. Yellowish sulfur granules cover these bacteria, which secrete enzymes to convert the granules to more soluble sulfate compounds. A wide variety of soil microbes then use the sulfates.

Still ponds or swamps that give off a rotten egg smell, characteristic of hydrogen sulfide, provide evidence of active sulfur cycling taking place underground. Because this cycle depends on anaerobic bacteria, it occurs in sediments and the deepest waters lacking oxygen. If a light were to be lowered into a swamp and clicked on, the sulfur bacteria would be green and pinkish-purple.

Aside from his studies on nitrogen and sulfur bacteria, Winogradsky examined the bacteria that generate energy by oxidizing or reducing iron. The iron cycle takes place in waters that drain from slow-moving bodies such as swamps and ponds and involves continual conversions of the element's chemical form by releasing or accepting electrons from other atoms. In soils suspected of having high iron levels, orange to reddish soil indicates that more oxidation (releasing electrons) is occurring than reduction (accepting electrons). Three prevalent bacteria that carry out this step are *Thiobacillus ferrooxidans*, which works in acidic conditions, *Gallionella* that prefers neutral conditions, and *Sulfolobus*, which grows best in acidic and high-temperature conditions. Microbiologists find *T. ferrooxidans* in areas that contain drainage from mining operations and *Sulfolobus* in sulfur hot springs. High-iron soils that are dark green or black contain more reduction than oxidation. The anaerobic *Geobacter*, *Desulfuromonas*, and *Ferribacterium* perform this reaction.

Winogradsky's legacy has been captured in a simple experiment that pulls together all of these metabolisms and mimics bacterial activities in nature. The "Winogradsky column" is a tall cylinder or jar filled with wet mud from a pond, lake, or ocean shore and topped with water. The amateur scientist adds shredded and chopped newspaper (as carbon source) and egg yolk (sulfur), and then puts the glass in a well-lighted place. After six weeks, bacteria from the mud settle into layers defined by oxygen levels with anaerobic mud below and aerated water above. The bacterial numbers start out low, but the appearance of colored striations in the column indicate the populations have grown to high densities. The colors give clues to the organization of bacteria in the column:

- Blue-green cyanobacteria receiving sunlight at the top
- Sulfide-using bacteria *Beggiatoa* and *Thiobacillus* in a light brown layer
- Photosynthetic *Rhodospirillum* in a large, nutrient-rich, rust-colored layer
- Red *Chromatium* in a low-oxygen layer using filtered light for photosynthesis
- Green *Chlorobium* absorbing hydrogen sulfide gas rising from the mud
- Brown anaerobic mud filled with hydrogen sulfide-producing *Desulfovibrio* and cellulose (newspaper)-degrading *Clostridium*

Iron-reducing bacteria, if present, live in the anaerobic sediment at the bottom of the column, and iron-oxidizing bacteria develop a rusty-red zone above the sediment.

In a single glass container, a person can watch the real activities of bacteria in nature. The Winogradsky column also creates a simplified microcosm of evolution; anaerobic actions beginning life in an oxygenless environment, and then progressing to photosynthesis and oxygen-respiring organisms.

Winogradsky columns can metabolize for months to years to decades. Microbiologists now build modified columns to emphasize a certain type of metabolism. For example, columns containing sediments from iron bogs or iron springs have more iron metabolism than a standard Winogradsky column.

## Feedback and ecosystem maintenance

Beijerinck and Winogradsky made a pivotal decision to study mixed populations of bacteria as they are in nature. By doing so, they helped define the concept of an ecosystem. The Winogradsky column contains numerous interdependencies between bacteria, yet it is a simple example of larger natural ecosystems.

A properly working ecosystem does not remain static but rather evolves in a process called succession. On a large scale, deforested land offers the best visible illustration of succession. New life takes hold on denuded land when cyanobacteria begin to grow in numbers. Some of the cyanobacteria team with new fungi entering the environment to form lichens that begin to cover the nutrient-scarce land. Mosses follow, and in turn small plants follow them. Over a period of months, higher plants such as bushes become established. Small trees and progressively larger, longer-lived trees establish over the next years. As this succession progresses, some species disappear as new more complex species emerge. Microscopic ecosystems follow a similar type of succession.

When bacteria enter a pristine habitat, nutrients may be plentiful and competition low. (Nature has no completely pristine habitats, but some natural events like floods and fires can create habitats that have lost a lot of their life and are ripe for recolonization.) The first bacteria to colonize the habitat are usually microbes that start out in the highest numbers or grow faster than the other microbes. Equally important, these bacteria have already adapted to the environmental conditions. They begin to change the habitat according to their specific type of metabolism. Some bacteria alter the pH, others remove all the oxygen, and some excrete simple organic compounds.

The altered conditions might favor another group of bacteria over the original species. For example, an acid-producing bacterium eventually chokes under the buildup of acids. But another species that uses organic acids as a carbon source sees the habitat as a nutrient-rich place to colonize. In rare instances, the original colonizer alters the environment so much that no other organism can live there. For example, areas exposed to mine drainage become increasingly acidic when *T. ferrooxidans* grows there and produces sulfuric acid as a by-product of the iron-sulfur compound pyrite. An ecosystem of diverse life cannot develop in situations like this, and the area turns into an extreme environment where only acid-loving extremophiles can live.

In the development of a healthy ecosystem, bacteria provide the foundation for food chains. Increasingly complex organisms become established. In healthy ecosystems, the new food chains develop associations that link them horizontally as well as vertically. In other words, a food web develops.

The more complex an ecosystem, the better it withstands changes in the environment. In simple ecosystems with few food chains all of the members depend on a relatively few species. If one or two species disappear, the entire ecosystem collapses. By contrast, complex ecosystems with many alternate paths for energy- and nutrient-sharing are versatile and can adjust to change. Rich biodiversity benefits all life, and this biodiversity extends all the way to microscopic life.

Ecosystems are hardwired to control the number and variety of species they contain. Two types of control processes operate: bottom-up and top-down. Bottom-up control uses microbes as the primary determinant of ecosystem health. If bacteria begin to disappear, the foundation of the ecosystem's food chains also disappears. Top-down ecosystem control theorizes that predators control the health of an ecosystem. By regulating the size of its prey, each member of an ecosystem prevents an exploding population of another member. Nature rarely follows hard and fast rules, so ecosystems tend to utilize a mixture of both control mechanisms.

In any ecosystem, organisms depend on feedback to help them regulate their activities. The simplest feedback mechanism to understand is food supply; when a person is full, he stops eating (hopefully). Being highly attuned to their environment, bacteria constantly interpret the immediate surroundings and respond by using feedback systems. For example, under starvation conditions, *Bacillus* turns into an endospore and myxobacteria produce fruiting bodies. When an ecosystem undergoes dramatic changes, even feedback may not be sufficient to save all of the system's members.

Microbial blooms are an example of an ecosystem gone out of balance. A bloom is a rapid overgrowth of microbes that drastically change an environment to the harm of other species. Blooms are caused by a sudden increase in numbers of aquatic algae, protozoa, or bacteria. Cyanobacteria and purple sulfur-metabolizing bacteria create most bacterial blooms, but bacteria play a part in algal blooms, too. Cyanobacteria and algae bloom in fresh and marine water when large, sudden influxes of nitrogen and/or phosphorus enter the environment. Runoff carrying fertilizer or manure from farmland acts as the main cause of blooms. Nitrogen and phosphorus wash into waters that usually contain low levels of these nutrients. The sudden bounty of nutrients causes an equally sudden explosion of microbes. As the microbial population grows increasingly dense, the cells release oxygen into the water as well as nutrients in the form of dead cells. Heterotrophic bacteria (bacteria that use sugars, fibers, amino acids, and fats) begin feasting, and they create a second bloom.

Bacteria in the second bloom are not photosynthetic so do not produce oxygen. Instead the fast-growing heterotrophs suck up all the oxygen from water around them. The oxygenless conditions soon cause other life to disappear; fish, crustaceans, and small invertebrates suffocate. Nutrient influx followed by ecosystem imbalance is called eutrophication. Cyanobacteria *Anabaena* and *Nostoc* are two common causes of blooms.

Cyanobacterial blooms now develop yearly in coastal areas and specific rivers worldwide and cause a health threat beyond the harm caused to aquatic life: cyanotoxins. Cyanotoxins are poisons released by cyanobacteria and that remain in the water after the bacteria subside. A serious occurrence of cyanotoxin contamination took place in Brazil in 1993. Fifty hospitalized dialysis patients died because their therapy used a water source contaminated with microcystin from the cyanobacterium *Microcystis*. (Water treatment technology has improved for removing pathogenic bacteria, but it remains poor at removing antibiotics, hormones, chemicals, and toxins.)

Anaerobic blooms occur when purple sulfur-metabolizing bacteria *Chromatium*, *Thiocapsa*, or *Thiospirillum* grow out of control. These blooms usually develop in oxygen-depleted waters in bogs and lagoons, leaving a telltale pink-purple sheen over the water.

Many blooms disappear on their own when seasons change and the hours of sunlight decrease, but several sites worldwide develop annual cyanobacterial blooms. Problem blooms return every year to the Great Lakes, the western United States, many Pacific islands, and lakes and rivers in Europe.

Lake blooms can also come from the anaerobic purple bacteria that live in darkness. Nutrient-rich lakes with a deep layer of bottom sediments give rise to large anaerobic populations that support communities of *Chromatium* and *Chlorobium* just above the sediment layer. These two species possess the unusual ability to catch filtered sunlight that penetrates past the photic zone. As anaerobes proliferate, they can turn the lake conditions unsuitable for other life.

In the 1970s Lake Císo in Spain became a topic for study because it had developed an anaerobic bloom of sulfur-metabolizing bacteria. The sediments emitted so much hydrogen sulfide that the gas filled the lake's entire water column, creating a rare type of anaerobic lake. Sulfate-rich water now runs into the bottom of the Lake Císo and water dense with bacteria flow from the upper layer. Most other anaerobic lakes have an upper layer of cyanobacteria, on top of a *Chromatium* layer, that sits atop much darker, sulfur-saturated water. Such an ecosystem is uninhabitable for fish and other animal life.

## Macrobiology

In optimal conditions, ecosystems do not go out of balance. Whether in lakes, soils, rumens, or insects, ecosystem members tend to self-regulate their populations. These ecosystems receive extensive research and usually become study models for microbiology students. Other ecosystems have offered few hints on how they work.

The luminescent bacterium *Vibrio phosphoreum* was discovered in the 1970s in specialized glands of certain deep sea organisms (lantern fish, angler fish, and some jellyfish and eel), and they have since been found in Alaska salmon. Their role in water ecology has puzzled scientists. The bacteria emit bluish-green light generated by the pigment luciferin, the same compound that lights fireflies and creates the nighttime phosphorescence sailors see in their ship's wake. Does *V. phosphoreum* benefit the fish or does the host benefit the bacterium? Perhaps neither organism cares about the other even though they live as a pair, a relationship called neutralism.

Microbial ecologists have barely scratched the surface of the relationships between bacteria and global ecology. Their challenges increase when considering almost inaccessible bacterial habitats deep in the Earth's mantle or miles under the ocean surface.

Only within the past decade or so have microbiologists extended the reach of their studies to depths of about two miles into the earth or approaching a mile into the polar ice sheets. The information ecologists draw upon to describe the functions of bacteria on Earth has come entirely from species close to or at the surface. Subsurface microbiology seeks to answer the questions of how bacteria of the deep contribute to life at the Earth's surface. What do these bacteria eat in the darkness? How do they relate to the evolution of life at the surface? Do they have any connection at all to life on other planets?

The U.S. Department of Energy launched a program on subsurface microbiology in 1986. Wells drilled to aquifers about 700 feet deep reached a population of diverse and novel bacteria. With the help of geologists and hydrologists, researchers either drilled to the depths or gained access to the deep subsurface via existing mines. As microbiologists probed deeper they found that bacteria became increasingly dependent on inorganic materials for survival and less on organic compounds.

Astrophysicist and NASA consultant Thomas Gold (who died in 2004) speculated in his book *The Deep Hot Biosphere* (1999) that the oceans' food chains begin not with microscopic marine life in the water, but deep in the Earth's lithosphere. These subsurface thermophiles, Gold proposed, exist on methane and hydrocarbons in massive untapped oil reserves and represent the closest relatives to life's ultimate ancestors. A controversy rumbles, distant from the daily concerns of most microbiologists, as to whether life emerged on Earth's surface or deep underground and grew outward to the surface. The bacteria that live in the deep-sea hydrothermal vents form the core of this argument because no one has as yet determined their origin.

A plan exists for building a subsurface physics laboratory in South Dakota's Homestake gold mine one and a half miles down. Geomicrobiologists, who study the interactions of microbes with geological formations, anxiously await. First, the construction must overcome issues of water purification, equipment installation, and the possibility of higher than normal radioactive bombardment. Then microbiologists will face the hurdle of growing these specialized bacteria under lab conditions.

Microbiologists continue to learn about the connection between the Earth's oil and subsurface bacteria. Some bacteria live in a world flooded with hydrocarbons, intense pressure at two and a half miles deep, and temperatures of 185°F: the world's oil reserves. The earliest studies on species recovered from the reserves revealed that many were related to surface species, surprising since the oil bacteria have been sealed off from other life for 200 to 500 million years. To defuse the inevitable charges of contamination that skeptics made toward this discovery, scientists have constructed small sampling capsules that open only when they reach oil and enclose their sample before returning to the surface.

A new science in oil microbiology has begun. Bacteria will play a pivotal role in oil refining, invention of fossil fuel alternatives, and oil spill cleanup. Microbiology has plans for the bacteria that live on oil. By analyzing the genes of bacteria recovered from oil and comparing them to genes in soil species at the surface, biologists may be able to locate new oil reserves. A similar array of genes between both groups could indicate that the surface microbes are living on oil seepage from oil reserves below them.

The relationship between oil and global ecology, or macrobiology, is complex. But at the core of oil's origin and its future sit the bacteria.

# 7

## Climate, bacteria, and a barrel of oil

A band of rock called the Isua formation runs along the edge of an inland ice cap from western Labrador to southwestern Greenland. The formation holds the oldest known rock found so far on Earth, dated at 3.8 billion years. Traces of fossilized life lace the Isua formation and analyses of its carbon content point to photosynthetic ancestors of cyanobacteria.

During the period in which the Isua formation developed, the Earth's atmosphere held no oxygen. Primitive photosynthetic microbes used the sun, carbon dioxide, and the Earth's elements (nitrogen, sulfur, phosphorus, salts, and metals) to sustain life. Their rudimentary photosynthetic reactions released little oxygen. Chemically unstable compounds in the atmosphere quickly captured what little oxygen the microbes liberated, and the oceans absorbed the rest. By 2.2 billion years ago, however, the oceans had accumulated enough dissolved oxygen to allow the gas to begin building up in the atmosphere. The oxygen levels in the atmosphere began to stabilize about 2 billion years ago.

Evolution is a change in an entire population due to small and discrete adaptations that favor species survival. The accumulation of oxygen on Earth signaled the development of a stable population of photosynthetic microbes that we now identify as primitive cyanobacteria. The cyanobacteria split into two evolutionary paths at least two billion years ago. One branch gave rise to plants. (Gene analysis suggests that the archaea branched off this path.) The second branch led to modern cyanobacteria and other bacteria.

By analyzing bacterial DNA, scientists have found that almost all bacteria contain DNA base sequences that are remnants of earlier evolutionary paths. In other words, bacteria have exchanged genes for so long that their evolution may resemble more of a network than a straight path. A few microbiologists have half-jokingly proposed that instead of estimating the thousands of bacterial species in the world, we should think of all bacteria as belonging to one giant species with a huge family tree of relatives.

The evolution of photosynthesis by whichever multiple paths it used accelerated the development of other biota. Microbial ecologist Patrick Jjemba has justifiably concluded, "The evolution of photosynthesis is the most important metabolic invention in the history of life on Earth." Bacterial diversity increased as oxygen levels rose from 0.1 percent (about 2.8 billion years ago) to 1 percent (2 billion years) to 10 percent (1.75 billion years). Not until the Cambrian Period, 543 to 490 million years ago, did oxygen reach its present concentration. The sudden increase in the diversity of life has prompted scientists to call it the Cambrian Explosion. The evolution of today's higher plants and animals took less time than the evolution of the Earth's first bacterial cell.

Although life developed into hundreds of millions of different forms, the variety of aerobic or anaerobic energy-production schemes inside cells remained disproportionately small. The pathway called glycolysis is life's universal pathway because it exists in every living thing. Bacteria use glycolysis as humans and other animals do by getting a small amount of energy from the breakdown of glucose to pyruvate. After glycolysis, various bacteria use a varied but limited choice of metabolisms. In addition to photosynthesis and glycolysis, bacteria use anaerobic fermentations, anaerobic or aerobic respiration, plus a small number of specialized metabolisms that branch off from these main metabolic pathways.

The story of oil began when oxygen accumulated in the atmosphere and fed the respiration of aerobic organisms. Food chains made of a widening diversity of life developed on continents and in the oceans. Bacteria, protozoa, algae, worms, and crustaceans built a hierarchy of prey and predators. The oceans took in dead bacteria, invertebrates, plankton, and the remains of prehistoric multicellular creatures. The majority of expired macro- and microscopic life never reached the ocean floor; other animals ate the organic matter as it sunk. But over millennia the populations of marine organisms increased, and more organic matter drifted down and built up in the sediments under the ocean.

The diversity of species that ended up in the organic sediments contributed to sediments' various forms of carbon. The Earth has been estimated to hold at present about 1.4 million known species and at least 10 times that number of undiscovered, uncharacterized species. Many times more species have already gone extinct than the number that survive today, yet today's biodiversity resulted directly from the Cambrian Explosion, the period in Earth's history when oxygen systems expanded faster than anaerobic systems.

## The story of oil

Plant and animal matter decomposed due to the action of bacteria millions of years ago as it does today. As each layer of organic matter under the ocean crushed the layers below, the pressure expelled water molecules. The sediments accumulated a dense mixture of carbon compounds, the majority of which were hydrocarbons, long carbon chains in which each carbon is saturated with hydrogen. Over millions of years, the pressure pushed the hydrocarbons deeper into the Earth and caused them to harden into a brownish-black solid. A chunk of this material viewed through a microscope would reveal fossilized bacteria—hence the name fossil fuel—with other bits of plant life, marine invertebrates, and shells.

Oil formation from the hard, black material required a precise combination of organic substances, pressure, time, and characteristics of the surrounding rock. Pressure from above pushed clumps of organic matter toward the Earth's center where it rose to about 180°F. Given enough time, the heating and the pressure turned the black rock into a liquid in which the hydrocarbon chains broke into a heterogeneous mixture of smaller chains. The pressure then pushed the liquid into pores in the surrounding rock. As the liquid squeezed through the network of pores, membrane constituents from bacteria mixed with it and increased its water-repelling property. The entire process produced crude oil.

At about 18,000 feet deep, oil remains in liquid form. Deeper, the intense pressure and heat further decompose the hydrocarbons into methane or natural gas. At shallower depths, the hydrocarbons remain solid and make up coal.

Microbiologists know that bacteria have played an integral part in formation of fossil fuels, but they still do not know all the ways in which bacterial metabolism altered oil's hydrocarbon composition from site to site. Oil shale, the rock that contains crude oil in a network of pores, contains chlorophyll pigments resembling those of modern photosynthetic bacteria. The late microbial ecologist Claude ZoBell has added that without bacteria oil would never have been formed. ZoBell theorized that subterranean bacteria acted on the longest hydrocarbons to make shorter (but still long) hydrocarbons. Crude oil contains hydrocarbon lengths from 8 to 80 carbons, and the relative composition varies between reserves and within the same reserve. For centuries, hydrocarbon-digesting anaerobic bacteria saturated the carbon atoms with hydrogen. These anaerobes also helped form natural gas, the very same methane produced inside ruminants and termites.

Fossil fuels can be viewed as renewable resources because the sedimentation process is continual. Organic matter continues to sink underfoot, and bacteria will eventually make more oil. But the process unfolds on a timescale that humanity does not comprehend.

Most people should by now understand that the rate of oil consumption outpaces the Earth's available oil. Saudi Arabian oil expert Sadad I. Al Husseini calculated in 2000 that the world's oil reserves would level off about the year 2004, and the plateau might last for no more than 15 years. After this plateau, the remaining oil becomes too difficult and/or too expensive to extract. The United States already passed this Rubicon in the early 1970s. A clue that signaled an oil deficit arose during that decade: increasing numbers of oil tankers bringing crude to the United States from halfway around the globe, often accompanied by spills.

Bacteria can help build second and third generations of alternative energies. Could bacteria be engineered to manufacture hydrocarbon fuels on a scale needed by the human population?

## Bacteria power

Unrefined crude oil poisons marine and terrestrial animals that ingest it. Oil's aromatic hydrocarbons, compounds with a carbon ring structure (benzene, toluene, xylene, and so on) damage tissue, enzymes, and nervous systems. Bacteria view crude oil as a carbon-rich and digestible food. Bioengineers have begun to turn to this process on its head.

Entrepreneurial companies such as LS9 in California have bioengineered *E. coli* and other bacteria to produce hydrocarbons that refineries can then turn into fuel without emitting sulfur gases made by conventional refineries. Microbiologists know how to make a minor alteration to bacterial fatty acid synthesis so that the cells produce gasoline instead of fats. Bioengineered species might soon churn out hydrocarbons of specific chain lengths as a way to adjust octane level.

The proportion of heavy, hard-to-extract oil has risen in oil reserves as Big Oil draws off the lighter, cleaner crude. Geomicrobiologists now search for bacteria that convert heavy oil to higher quality fuel for combustion engines. Increasing today's oil recovery rate by as little as 5 percent would make a substantive impact on world oil supply.

The bacteria that fix nitrogen, that is, capture nitrogen gas directly from the air, release hydrogen gas, which has been touted as an alternative to fossil fuel. Heterotrophs, some photosynthetic bacteria, and anaerobes make hydrogen as part of their normal metabolism. Bacterial hydrogen production for future fuels would require large fermenters designed to allow in sunlight for photosynthesis and possibly more than one species working in concert. For example an anaerobe that produces hydrogen might pair with anaerobic photosynthetizers that energize the system by absorbing sunlight.

Current chemical methods for making hydrogen involve breaking apart water molecules in a costly and technologically challenging process. Bacteria use the enzyme hydrogenase to split water into hydrogen and oxygen with less energy demand than the same reaction in a manufacturing plant. Some bacterial hydrogenases need only a small supply of selenium, iron, and nickel to stabilize the reaction. Biochemists are already working on a thermophile *Clostridium* that performs the reaction at about 140°F and dispenses with the need for added metals.

Oxford University chemists have also attached hydrogenase and a light-sensitive dye to microscopic titanium dioxide beads. In this system, photosynthetic microbes supply their own energy by capturing solar energy. No conventional chemical companies can make the same claim.

Similar nonsunlit systems include *E. coli* hydrogenase with the enzyme carbon monoxide dehydrogenase (CMD) from *Carboxydothermus hydrogenoformans*. CMD splits carbon monoxide. The overall reaction is

$$\text{carbon monoxide (CO)} + \text{water } (H_2O) \rightarrow \text{carbon dioxide } (CO_2) + \text{hydrogen } (H_2)$$

*C. hydrogenoformans* that catalyzes this reaction is an anaerobe isolated in 1991 from a freshwater hot spring on Kunashir Island in the Sea of Japan. The German Collection of Microorganisms in Braunschweig owns one of the world's few cultures of this obscure microbe.

Astute readers will notice that the preceding reaction gets rid of the greenhouse gas carbon monoxide but produces another culprit in global warming, carbon dioxide. Scientists have dreamed up various ways to pull carbon dioxide from the air. Ideas include giant filters strewn across the landscape to suck in carbon dioxide and then pump it deep into the earth. Others have proposed seeding the oceans with nutrients so that algae and cyanobacteria increase and thus consume more carbon dioxide.

Carbon-dioxide consumers in the bacterial world occupy special niches. Chemolithotrophs (growing only on inorganic salts and carbon dioxide) and photolithotrophs (growing on sunlight and carbon dioxide) draw some of this gas from the atmosphere. The other important consumers of carbon dioxide live in dark places where they digest organic matter and prevent the Earth from becoming choked in waste.

## How is a cow like a cockroach?

The methane gas emanating from sewage treatment plants, landfills, and the muck submerged in swamps comes mainly from methane-producing archaea. These so-called methanogens interact with bacteria in a way that allows both to thrive and keep ecosystems running.

Methanogens sustain cattle, goats, sheep, deer, elephants, and all other ruminants plus cockroaches, termites, beetles, and millipedes—thousands of arthropod species in all. Their digestive tracts contain a heterogeneous mixture of microbes from the three domains of living things: Archaea, Bacteria, and Eukarya. The bacteria and archaea cling to the rumen wall and to feedstuffs entering the rumen while protozoa tend to stay in the liquid.

A cow's four-part digestive organ—rumen, reticulum, omasum, and abomasum—evolved for fermentation. Ruminant animals scarcely chew their food after yanking it from soil; they masticate just enough to mix the grasses with saliva, and then send the bolus into the esophagus leading to the rumen. The cow rumen holds up to 20 gallons, the interior resembling a perpetual washing machine lined with small protuberances called papillae. These structures increase the rumen's inner surface area to make absorption more efficient and to increase attachment sites for microbes. Rumen fluid ranges from Kelly green from grass diets to olive-green when the cow gets mainly a hay diet. Every minute or so the esophagus launches a bolus into the mix like a torpedo. The fluid softens the bolus, and then the animal regurgitates and rechews it. After "chewing the cud," the bolus goes back to the rumen where bacteria and protozoa continue the fiber digestion. As the rumen contents slosh about, the animal regurgitates larger pieces and sends smaller denser pieces to the intestines where a different population of bacteria continues the digestion.

A microscope slide holding a drop of rumen fluid or a speck from a cockroach's innards reveals a mob of microbial life. Cocci and rods bob in the currents. Every second or so a spirillum twirls through the microscope's field; blink and you have missed it. Protozoa come and go, looking massive next to the bacteria. These eukaryotes range from 20 to more than 100 times the volume of bacteria. Some flagellated protozoa poke through the liquid in fits and starts while other protozoa blanketed in cilia whiz past.

Most of the digestive anaerobes differ from *E. coli* because they cannot tolerate even miniscule amounts of oxygen. These überanaerobes (obligate anaerobes to microbiologists) include *Bacteroides*, *Butyrivibrio*, *Clostridium*, *Eubacterium*, *Lactobacillus*, *Peptostreptococcus*, *Ruminococcus*, *Selenomonas*, *Streptococcus*,

*Succinimonas*, *Succinivibrio*, and *Veillonella*. Cattle also harbor *Lactobacillis*, *Clostridium*, and *E. coli*—cattle is humans' main source of the deadly *E. coli* O157 when farm waste contaminates food. At least 20 species of archaea and 50 species of protozoa also inhabit the digestive tract.

When fibers (cellulose, hemicelluloses) and polysaccharides enter the rumen, bacteria degrade these large compounds into smaller sugars for energy. Protozoa subsist on sugars but also graze on the variety of bacteria and archaea. The archaeal methanogens use carbon dioxide plus vitamins and minerals available in the rumen fluid.

The cow uses relatively little of the nutrients in grass and grain directly. Ruminants live mostly on the volatile compounds emitted by the bacteria. These so-called volatile fatty acids (VFAs) named acetic (two carbons), propionic (three), and butyric (four) acids pass through the animal's gut lining and enter the bloodstream. The fat and flavor of fresh cow's milk result from the mammary gland's synthesis of long fats from the short VFAs. Goats produce a different array of fats from the same three VFAs, which results in distinctive flavors in products made from goat's milk.

Cows receive most of their amino acids and vitamins from bacteria that the animal's digestive enzymes degrade. Unlike humans, ruminants survive on very poor quality protein, meaning the protein contains a limited variety of amino acids, because the bacteria improve the variety of amino acids available for absorption.

Cattle spend one-third of their time eating, one-third ruminating or chewing the cud, and one-third resting. During rest, bacterial activity reaches its peak: Large molecules decompose in fermentation to VFAs, carbon dioxide, and a little hydrogen. These reactions would soon stop if carbon dioxide built up in the gut. Methanogenic archaea play the vital role of absorbing carbon dioxide as it appears and turning it into methane:

$$CO_2 + H_2 \rightarrow CH_4$$

A dairy cow with a 15-gallon rumen belches 65 to 130 gallons or 5,370 to 10,740 cubic feet of methane a day. The world's domesticated and wild ruminants produce about 22 percent of the atmosphere's methane, one million tons of methane put into the atmosphere a year.

Since methane exerts more than 20 times the atmosphere-warming effect of carbon dioxide, ruminants contribute to global warming. When the media make coy references to ruminant flatulence as a major cause of global warming they really should blame belching.

Microbiologists study the goings on inside cow rumens by using fistulated animals. A fistula is an opening about the diameter of an orange leading from the outside of the animal to the inside of the rumen. The left wall of a cow's rumen lies against the animal's left side, making the distance from outside to inside less than 3 inches. After a veterinarian surgically fistulates the left side of the animal, the patient recovers quickly and begins eating again within the first few hours after surgery. (Humans can last days without food, but a ruminant cannot go 24 hours without food before becoming deathly ill.) The fistula, like a rubberized doughnut, can be closed with a tight-fitting plastic plug. When opening a fistula plug, a rush of methane bursts from within.

Cockroaches use processes similar to ruminants but with a more active role by protozoa. Bacteria and archaea living inside the protozoa that live inside the insect's gut carry out the chemical reactions of digestion. Protozoa presumably take in nutrients that sustain the prokaryotes and protect them from predation by other protozoa. As a result, 80 percent of the (American) cockroach's methane emissions comes from its protozoa.

The protozoa inside ruminants, cockroaches, and termites live in mutualistic symbiosis with the host. Termites contain symbionts within symbionts. The insect lacks fiber-degrading enzymes, so it depends on gut protozoa to digest the wood fibers. But the protozoa, such as *Trichonympha sphaerica*, also make little progress in digesting wood. *T. sphaerica* relies on spirochete (spiral-shaped) bacteria living inside it. The bacteria produce the enzyme cellulase that decomposes the cellulose so that the insect, the protozoa, and the bacteria all benefit.

A second group of bacteria live on the outside of termite protozoa. Some spirochetes and other rod-shaped cells line up in precise rows in grooves between the protozoan's cilia. Electron microscopy has revealed that the curvy spirochetes line up end-to-end and undulate in unison. The protozoan moves by the combined action of its cilia and the coordinated beating of thousands of spirochete flagella, creating a smooth wave of propulsion. No one has yet figured out if the protozoa tell the bacteria where to swim or if the bacteria control where protozoa go. Regardless of the answer, protozoa need their bacteria; if the bacteria disappear, the protozoan stops dead in the water.

## Microscopic power plants

In the 1990s, Al Gore's tireless campaign to address global warming prompted scientists to identify the world's main sources of methane. Twenty times more active in warming the atmosphere than carbon dioxide, methane became a strategic target in the global warming campaign. The scientists estimated that enteric fermentations of ruminants and insects account for almost 25 percent of the atmosphere's methane. Cattle manure accounts for another 7.5 percent.

More than half of the methane from human-made structures such as landfills and wastewater treatment already goes into systems that use it as an energy source. The methane from swamps, stagnant ponds, manure piles, and domesticated and wild ruminants goes lost to the atmosphere. An adult cow produces about 27 pounds of solid waste daily and the 100 million cattle in the United States add close to 14,000 tons of manure to waste piles every day. Central Vermont Public Service offers manure-derived methane, or "cow power," to more than 3,000 homes and businesses. The state's dairy farms supply the manure that produces the biogas, and the utility converts the gaseous energy to electrical energy and distributes the electricity.

Bacteria in nature or in test tubes always take the most efficient path for finding, absorbing, and metabolizing nutrients. Heterotrophs prefer sugars, fibers, amino acids, and fats for energy and building new cells. Other bacteria called autotrophs thrive on a less heterogeneous variety of nutrients, namely water and carbon dioxide for cell-building and sunlight or metal for energy. Autotrophs (also called lithotrophs) grow on a chunk of rock devoid of organic matter or in the nutrient-empty ultrapure water used in semiconductor manufacturing. The bacteria being discovered in subsurface microbiology are all autotrophs. They get small bits of energy from chemical reactions between water and basalt, and scavenge nitrogen and sulfur from tiny pockets of air.

Heterotroph and autotroph energy production happens in the cell membrane, a multilayered covering that lies just inside the cell wall. Energy generation in bacteria resembles that of humans in that it uses a stepwise transfer of electrons from compound to compound. Each transfer produces small spurts of energy. Humans use membrane-bound proteins called cytochromes to perform most of the electron transfers. Bacteria depend on pigments. The blue-green hues of ocean and freshwater cyanobacteria, the striking colors in hot springs from sulfur and iron metabolizers, and the green and purple intertidal flats populated by photosynthetic bacteria all give evidence that bacteria are hard at work.

Bacteria can be harnessed to produce energy directly rather than energy in the form of fuel. University of Massachusetts microbiologists Derek Lovley and Gemma Reguera have showed that biofilms grow tiny filaments between cells. These filaments act as "nano-wires" to transmit electrical current, which the cell consortium amplifies about tenfold when electrons travel through the film. Perhaps energy companies will one day feed sugar and oxygen to massive biofilm fields and thus produce electricity as well as clean water, a by-product of photosynthesis. Algae and cyanobacteria both possess the capabilities to do this. Biotechnology could also engineer the microbes to produce hydrogen or ethanol.

## The waste problem

Bacteria degrade pollutants in soil, surface waters, and groundwaters. These pollutants include pesticides, vehicle and jet fuels, paints, organic solvents, and buried ammunition. Bioremediation scientists take genes from bacteria that metabolize these pollutants and insert the genes into bacteria that grow faster in nature. Bioremediation laboratories now have collections of bacteria that degrade chemical pollutants, detoxify metals such as mercury, or decompose radioactive compounds. Specialized bioreactors can grow biofilms on their interior surfaces (see Figure 7.1) and remove pollutants from water as it flows through the vessel. Bioremediation also seeks the bacteria that live in the acid drainage from ore and coal mines and which trickles 24/7 into more than 10,000 miles of U.S. rivers and streams. The traits enabling the bacteria to thrive in these places make them perfect gene donors for making bioengineered bacteria for mining pollution cleanup.

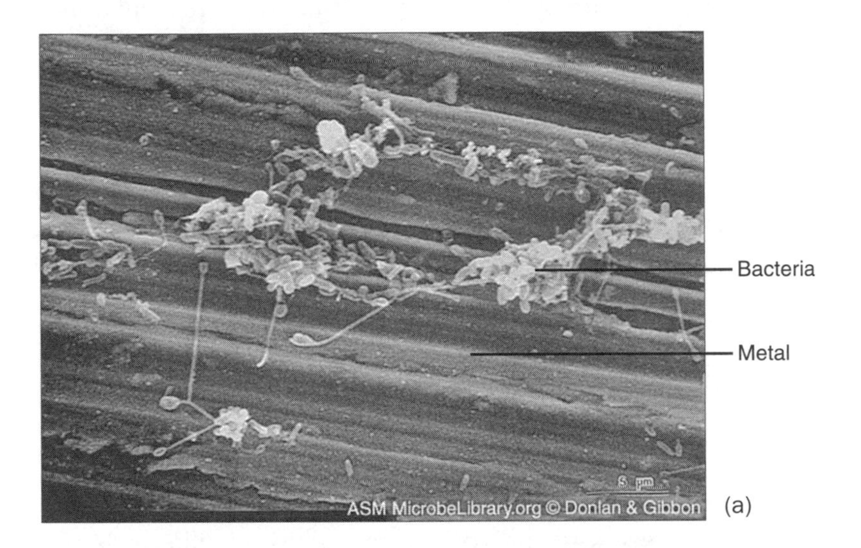


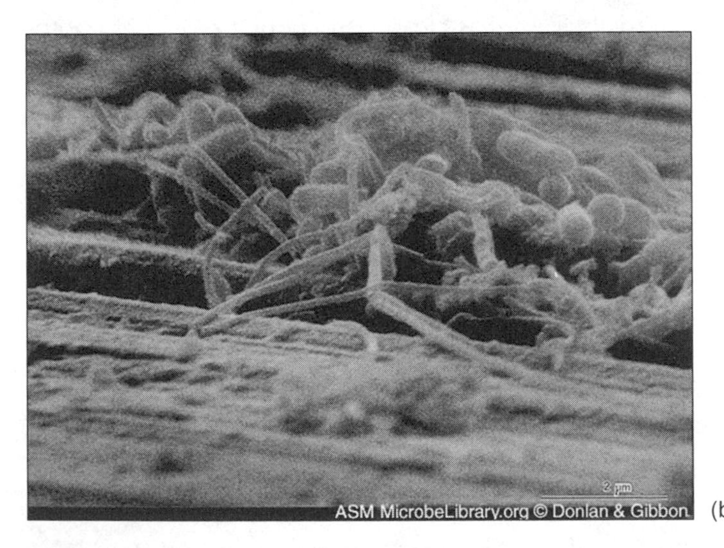


Figure 7.1 Community formation. In this series of photos, bacteria colonize the metal surface of a cooling system condenser, a process called fouling. (a) Scattered bacteria adhere to a copper-nickel surface; (b) cells and extracellular materials accumulate;

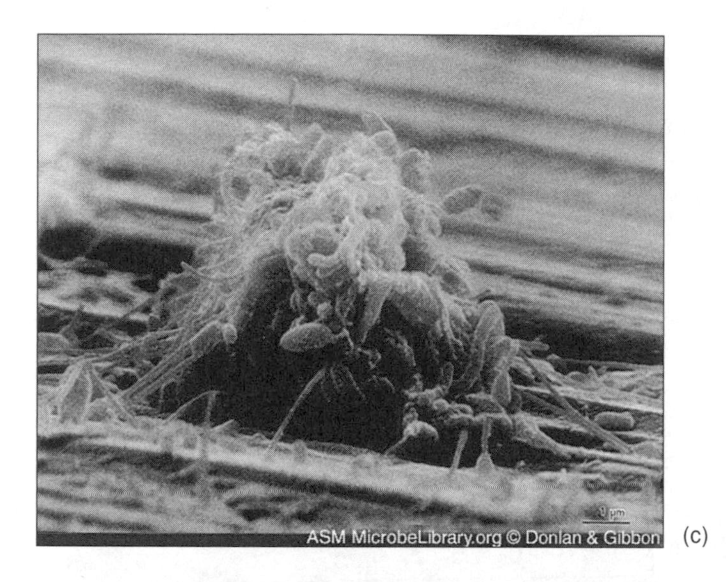


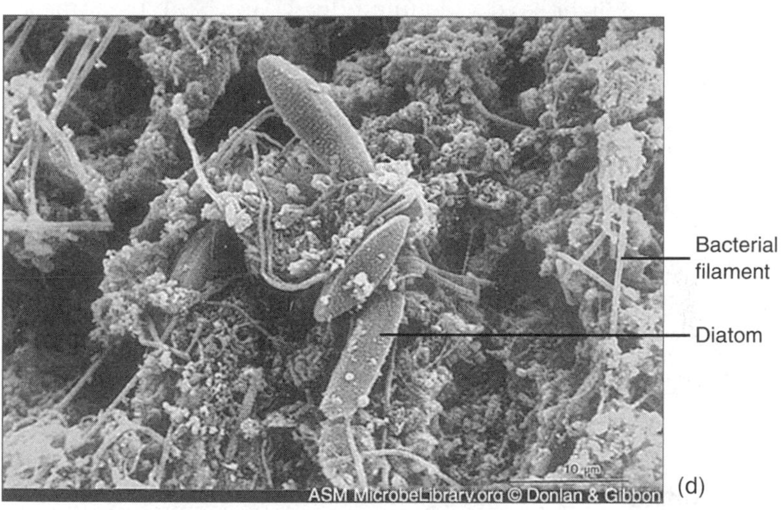


Figure 7.1 (c) filaments extend and trap more cells; and (d) bacteria, freshwater diatoms (type of algae), corrosion products, and clay particles imbed in intertwining filaments. (Reproduced with permission of the American Society for Microbiology MicrobeLibrary (http://www.microbelibrary.org))

Wastewater treatment plants rely on mixtures of aerobic bacteria to degrade the substances in the incoming water. This step happens in the plant's large outdoor pools filled with dark liquid. Wastewater plants mix the suspension with big paddles and constantly bubble air through the water to keep the bacteria growing. Treated wastewater gets a dose of chlorine to kill both pathogens and good bacteria before returning to the environment. The heavy sludge that settles to the bottom of wastewater pools receives extra digestion in closed tanks filled with anaerobic bacteria.

Wastewater treatment anaerobes break down tough substances like plant fibers and paper, but like rumen bacteria they emit methane gas. For many years, wastewater treatment plants burned off the methane stream from sludge digester tanks. Most now capture the gas and burn it for energy.

Methane production is a double-edged sword, a free energy source and a greenhouse gas. When too much methane enters the atmosphere, it can be viewed as a type of waste like carbon monoxide and hydrogen sulfides. A group of bacteria called methanotrophs use methane for both carbon and energy and thus remove some of the world's excess methane. *Methylobacteria*, *Methylococcus*, and *Xanthobacter* live in many of the same places as methanogens and absorb the gas as it is produced. For example, still, swampy water contains small bubbles of methane from the methanogens living in the anaerobic sediments at the bottom. Methanotrophs live just above this region, capturing some of the methane as it rises. *Xanthobacter* possesses an additional ability to combine oxygen with hydrogen gas, also made by methanogens, in an explosive reaction called the knallgas reaction ($O_2 + 2\ H_2 \rightarrow H_2O$). Knallgas bacteria have systems that control this reaction to make energy for the cell without blowing up!

Methanotrophs use the enzyme methane monooxygenase in its metabolism. This enzyme also breaks down a toxic chlorinated solvent called trichloroethylene (TCE). TCE is a pollutant in soil and groundwaters and harms almost every system in the body. Electroplating, metal degreasing, semiconductor manufacture, steel and rubber manufacturing, pulp and paper operations, and dry cleaning use TCE. Methanotrophs may soon become a tool for cleaning TCE out of contaminated aquifers and land. If microbiologists do put methanotrophs to work in labs, they will avoid *Xanthobacter*; knallgas bacteria require the explosive mixture of hydrogen and oxygen.

*Thiobacillus ferrooxidans* presents an equally important but more complex involvement with pollution. *T. ferrooxidans* thrives in very acidic conditions and gets energy from inorganic iron- and sulfur-containing compounds. All of these features occur in mine tailings, the fluids that flow from ore and coal mines. Mine tailings cause considerable damage to stream and river ecosystems. *T. ferrooxidans* reacts with iron pyrite to make tailings even more acidic and caustic to the environment.

Mining site remediation currently uses chemicals to absorb or neutralize the acid, but sulfate-reducing bacteria offer an option because they alter the acid-producing reactions of *T. ferrooxidans*. Sulfate-reducing bacteria begin with the prefix "Desulfo-"—such as, *Desulfococcus*, *Desulfovibrio*, and *Desulfobacter*.

Despite *T. ferrooxidans*'s penchant for making a bad environmental situation worse, the species has been put to good use too. *T. ferrooxidans* recovers metal from ore deposits and also reduces sulfur in coal, a step in making low-sulfur or "clean coal." Conventional coal burning releases the greenhouse gas sulfur dioxide from pyrite and contributes to sulfuric acid formation in the atmosphere, the cause of acid rain.

Depletion of high-grade metal ores in the United States has made the recovery of low-grade ores critical to the metals industry. But high cost prevents the extraction of metals from low-grade ores by the usual smelting process. *T. ferrooxidans* and the similar species *T. thiooxidans* extract metals such as copper and uranium from ores filled with iron and sulfur compounds. For example, either of these species can recover copper from the copper-iron-sulfur compound chalcopyrite. The bacteria do the recovery exclusively with enzymes, an example of white biotechnology. Bacterial bioleaching also recovers some of the iron along with copper, and both are recycled by the metals industry into new products.

Similar mechanisms have been used to extract uranium from gold ore. Bioleaching can recover 90 percent of the desired metal from low-grade ores and saves on the high energy cost of smelting.

## Bacteria on Mars

The ability of bacteria such as *T. ferrooxidans* to live surrounded by caustic chemicals or subsurface bacteria to eat rock keeps alive the notion that bacteria might live on other planets. Life on Mars represents more than an academic pursuit. Enzymes produced by potential Martian bacteria might have superior faculties to recover metals or make energy from Earth's greenhouse gases. Mars' atmosphere contains more than 95 percent carbon dioxide, about 3 percent nitrogen, and lesser amounts of oxygen, argon, and carbon monoxide. Other than argon, bacteria on Earth use all of these gases. Earth's autotrophs live on energy sources much like the elements on Mars, that is, silicon, iron, magnesium, calcium, sulfur, aluminum, potassium, sodium, and chlorine.

Recently microbiologists have developed a theory that the Earth's earliest bacteria may have degraded rock and thus carved out miniature caves that may have served as protective habitats. Earth's atmosphere 2.75 billion years ago, the estimated age of the caves, held no ozone layer to protect the breakdown of macromolecules by ultraviolet radiation. The caves would have shielded bacteria, protected their DNA from destruction, and provided a probable site for water condensation. The ancient anaerobes possibly used the minerals in the rock to which they attached and absorbed methane percolating up from sediments. This scenario does not seem implausible knowing what we do about extremophiles.

Birger Rasmussen of John Curtin University in Australia has begun a global discussion on whether cave-dwelling bacteria indicate similar bacteria might live on Mars. By analyzing the chemistry of ancient microbial deposits attached to the cave ceilings, he has presumed that the early bacteria used sulfur and methane, probably had access to water, and likely lived in a biofilm community.

Many scientists have been unwilling to make the giant leap from bacteria on Earth to bacteria on other planets. Assuming extraterrestrial life follows the same biochemical principles as on Earth, bacteria on Mars would probably exist in the subsurface for the same reasons cave dwellers developed.

Theories abound on the possibilities of life on other planets in distant solar systems or on Mars. The three main themes of interplanetary under study in astrobiology are water, methane, and minerals.

In 1996, NASA added fuel to the "bacteria on Mars" fire when it announced that a meteorite that had crashed near Antarctica 13,000 years ago held traces of bacterial growth. Scientists named the meteorite ALH 84001 and recovered it in 1984. By the early 1990s, NASA scientists believed it had blasted off from Mars and traveled interplanetary space for 16 billion years. Astrobiologists, meanwhile, focused on tiny wormlike structures embedded in the rock that resembled fossilized microbes. Analysis of the structures' elements suggested that the structures were more of biological origin than geological. With prior discovery of ancient rivers and seas on Mars' surface, science seemed to hold circumstantial evidence of water and life on the Red Planet.

The analysis of Mars' atmosphere has also provided evidence of methane. Considering that Earth's methane is almost 95 percent of biological origin, the presence of this gas on Mars has been viewed by some astrobiologists as another point in favor of life on Mars. Earth's atmospheric methane on a volume basis is 1,750 parts per billion, but that of Mars is only 10 parts per billion. No one knows why the disparity exists between the two planets or if any of Mars' methane came from living things.

Another research team examined the meteorite's mineral content and found magnetite crystals similar to the magnetosomes in Earth's *Aquaspirllium magnetotacticum*. Dennis Bazylinsky of the University of Nevada-Las Vegas has been studying magnetotactic bacteria for more than 20 years. When reviewing the meteorite data, he found the magnetic crystals in the meteorite to be identical to the crystals made by Earth's magnetic bacteria. Again, scientists held circumstantial evidence in their hands, but the task of comparing Earth's magnetic bacteria to extraterrestrial crystals would not be easy. Very few cultures of magnetic bacteria exist in laboratories worldwide. New strains are known to exist in nature but they live in difficult-to-reach marine sediments.

Naysayers to life on Mars have pointed out that methane and inorganic structures may resemble conditions on Earth but can also be explained in nonbiological terms, which is true. Microbiologists have questioned the meteorite's "worm holes" because they are much smaller than the smallest Earth bacteria and thus unlikely to contain all the molecules needed for life. Of course, those scientists are speaking about Earth life, which may be a bit egocentric considering the size of the universe. By 2000, however, most astrobiologists concluded that the worm holes were probably fossilized debris, organic debris perhaps, but not microbial.

Conclusive studies on nanobacteria on Earth may recharge the life-on-Mars debate. Finnish researcher Olavi Kajander discovered nanobacteria in 1988, but the majority of microbiologists rejected the idea of their existence. (Notice how every new discovery mentioned in this book endured a period of denial?) More than a decade of study on nanobacteria suggested that these microbes played a pathogenic role in arterial and kidney calcification.

By 2005, literature had accumulated on *Nanobacterium sanguineum*, a gram-negative motile species with a calcium-coated outer shell. The bacterium measures only 20 to 200 nm across, but it is big enough to contain 16S rRNA. Studies on *N. sanguineum* have followed a similar path as studies in the Golden Age of Microbiology: The medical importance of the microbe has superseded environmental studies. But nanobiology will very soon be part of the growing science in interplanetary biology.

## Shaping the planet

The Earth's biosphere consists of millions of ecosystems. When ecosystems interrelate, they form large ecosystem communities. The Earth thus has grassland, rainforest, polar communities, and so on. Although members interact at boundaries called edges, such as the interface between marine and shore life, many of Earth's communities remain separated by distance. Migrating herds and birds connect some communities but not all of them at once. Only bacteria connect all of Earth's communities by the constant recycling of nutrients through soil, the oceans, and the atmosphere.

No one needs a degree in microbiology to find these microbes all around and performing their life-giving activities. In the country, notice the lichen growing on rocks, dead leaves decomposing underfoot, and the glimpses of color when a lake ripples. If you live in the city, bacteria live all around. Biofilms coat storm drains and metal-metabolizing bacteria weather bridges and buildings. Soot in the air carries bacteria from block to block. It is easier to detect the invisible universe than it is to find places having no bacteria.

You may never look at your surroundings the same way again, and that is a good thing. Appreciating bacteria is the best way to acknowledge the larger community of Earth. When I was in college in the 1970s, I realized microbiology is a hard subject. It encompasses the basics of cell biology, covers chemical and biochemical reactions, touches on the Earth sciences, and is intimately connected to genetics. Microbiology recruits only scientists willing to study organisms they cannot see. But it is impossible to delve deeper into the microbial world without seeing that bacteria run this planet. Humans reap the benefits of bacterial actions when they discard garbage, avoid infection (remember the skin's good bacteria), and simply breathe.

Bacteria should not be synonymous with disease. Making cheese out of milk also seems to sell these microbes short. Because of bacteria, our lives are richer, healthier, and more hopeful. Hopeful because no matter what predicament humanity puts itself in, there is a very good chance that a bacterium somewhere can solve the problem.

Stop worrying about germs and start appreciating bacteria. Few pathogenic bacteria exist that cannot be stopped by simply washing hands, preparing food properly, and steering clear of others who are obviously sick. As for the good bacteria that fill the environment, we need not nurture them because they grow just fine without any help from humans. In the process, bacteria supply us with the nutrients we need. Bacteria shape the planet and they also shape us.

For safety's sake, thinking of bacteria as occasional enemies as well as constant allies helps maintain your health. In the bigger picture, however, bacteria are your best friends. They welcomed humanity into their home tens of thousands of years ago, and they will stay with you to the end. Bacteria work behind the scenes to protect us, feed us, and decompose our wastes. I cannot think of a better ally than bacteria.

# Epilogue: How microbiologists grow bacteria

Spending a career working with invisible objects can stretch a person's patience. Bacteria demand that a microbiologist wait several hours, overnight, or even several days before multiplying to high numbers. *Mycobacterium* takes three weeks to reach numbers high enough to study the organism. Finally, a person cannot call herself a microbiologist without mastering the art of aseptic technique. The technique truly is an art because no two bacterial cultures behave exactly the same way, and the avenues for contamination seem limitless. The standard practices described here avoid some of the pitfalls that students new to microbiology see when growing bacteria.

Microbiology samples may be patient specimens (blood, sputum, stool, and so on), foods, consumer products, soil, drinking water, untreated surface waters, or wastewater. Microbiologists usually take samples of 100 milliliters liquid or 10 grams of a solid to a laboratory for "processing," which is the series of steps needed to determine if bacteria are in the sample, how many, and what kinds.

Microbiology employs two aids for working with the huge numbers common in this field. First, microbiologists dilute samples containing millions or billions of cells in a technique called serial dilution. Second, the microbiologist converts these large numbers to logarithms.

## Serial dilution

A sample containing more than a million bacteria per milliliter or gram is too concentrated with cells for scientists to study a species and draw meaningful conclusions. Rather than juggle numbers of this magnitude, microbiologists dilute each sample sequentially to reduce the cell concentration to between 30 and 300 cells per milliliter.

The method called serial dilution consists of a set of tubes each containing 9.0 milliliters of sterile buffer (water with a small amount of salts to maintain a constant pH range). By taking one milliliter of the sample and adding it to one of the 9-milliliter tubes, the microbiologist has made a one-to-ten dilution, or 1:10. With this step, a sample containing three million cells per milliliter now contains 300,000. A milliliter of this new dilution transferred to another sterile 9-milliliter volume lowers the concentration to 30,000 cells per milliliter. The microbiologist continues diluting each new dilution until arriving at what he assumes is a much lower concentration than the original sample. This is tricky because the process is done on supposition. When microbiologists receive a sample from a patient, food, or the environment, they have no idea if the sample contains millions of bacteria or a few. Serial dilution helps span the range of possible concentrations to determine the actual concentration of cells in a sample.

Following the serial dilution, the microbiologist has a set of tubes before him, each tube containing one-tenth as many cells as the preceding tube. The next step involves inoculating agar plates with small volumes, called aliquots, from each dilution. The microbiologist might take a 0.1-milliliter aliquot from each dilution and put this amount onto individual sterile agar plates. For example, 0.1 milliliter of the 1:10 dilution goes onto a plate (microbiologists usually include duplicate or triplicate plates for each dilution), 0.1 milliliter of the 1:100 dilution does the same, and so on. After all of these transfers have been completed, the microbiologist has a set of inoculated plates each containing a subsample (the aliquot) from the 1:10, 1:100, 1:1,000, 1:10,000, and 1:100,000 dilutions.

Next, the microbiologist spreads each of the aliquots over the agar surface to spread out whatever bacteria may be there—remember, they are invisible. This spreading step requires a sterile glass or plastic rod about seven inches long with a bend at one end about an inch from the end of the rod. Visualize a hockey stick shape. These spreaders are in fact called "hockey sticks" by microbiologists. When the aliquot has been spread as a thin transparent film over the agar surface, the agar is called a spread plate. Each plate comes with a cover, which now goes onto the inoculated spread plate.

The scientist puts the entire stack of spread plates into an incubator set at a favorable temperature. Although a stack of agar plates in an incubator seems an obvious space-saving arrangement, this innovation of German bacteriologist J. R. Petri in 1887 changed microbiology. The stackable, compact Petri dishes enabled microbiologists to study more replicates and a wider variety of microbes than in previous experiments.

Most bacteria recovered from temperate environments grow at body temperature, so incubators can be set to about 98.6°F (37°C) for the incubation step. Many foodborne contaminants and almost all pathogens and native flora prefer this temperature. Soil and water microbes and some foodborne psychrophiles grow better at lower temperatures.

Incubation lasts overnight, a day or two, or several days to weeks, depending on the bacterium. After incubation of the plates, the microbiologist sees visible colonies, usually no bigger than one-eighth of an inch in diameter, each containing millions of bacteria.

## Counting bacteria

A colony of bacteria growing on agar contains identical cells that have all descended from a single ancestor cell. When a microbiologist inoculates agar, individual bacteria disperse in the medium. During incubation, each cell from the inoculum doubles in number every half hour or so, depending on species, until they form the visible mass of cells known as a colony. Microbiologists call the colonies CFUs for colony-forming units and count them either manually under a magnifying glass or electronically by scanning the agar plate with a laser beam.

Samples containing several thousand to millions of cells would create an almost contiguous sheet of colonies unless the microbiologist serially dilutes the sample before inoculating the plates. Serial dilution produces plates containing between 30 and 300 CFUs, most of which are spatially separated from each other and easy to count. Microbiologists prefer plates with this many colonies because CFU numbers of less than 30 do not give consistently accurate results, and plates with 300 or more colonies are too dense to count. On densely populated plates, bacteria begin inhibiting the growth of nearby colonies by using up nutrients and excreting antimicrobial substances.

To determine the number of bacteria in a liquid culture, the microbiologist selects duplicate plates containing 30 to 300 colonies each. In this example, the plates that had been inoculated with 0.1 milliliter of the 1:10,000 dilution look like they have between 30 and 300 colonies. After counting the number of CFUs on each duplicate plate, the microbiologist discovers one plate has 98 colonies and the second has 138 colonies. The average of the two plate counts equals 118. Now the microbiologist must account for the dilutions to calculate the number of bacteria that were in the original sample.

In the first step, the microbiologist multiplies 118 by the dilution, in this case, 1:10,000:

$$118 \times 10{,}000 = 1{,}180{,}000 \text{ or } 1.18 \times 10^6$$

The aliquot volume was only 0.1 milliliter, which is equivalent to diluting a milliliter by 1:10. To correct for this dilution, the microbiologist multiplies the above result by 10:

$$10 \times 1{,}180{,}000 = 11{,}800{,}000 \text{ or } 1.18 \times 10^7$$

The original culture therefore held almost 12 billion bacteria. In microbiology, such large microbial numbers occur often. Soil, marine water, surface freshwaters, and the animal digestive tract all contain similar high bacterial concentrations.

## Logarithms

Numbers of several million or billion can be unwieldy for calculations. Furthermore, when a number as large as $1.18 \times 10^6$ is doubled to $2.36 \times 10^6$ or even tripled, the differences between these numbers are not meaningful in microbiology. Variability in nature can cause replicate cultures prepared exactly the same way to produce different concentrations of bacteria. Microbiologists therefore use logarithms to make very large numbers easier to use in calculations and to help discern significant differences between large numbers.

Understanding the definition of a logarithm (abbreviated to log) can be difficult, but an example helps. For the number $1.0 \times 10^5$, the log is 5.00. The log for $1.0 \times 10^6$ equals 6.00. Numbers that fall in between whole numbers also can be converted to a log value. For example, the log of $5.0 \times 10^5$ equals 5.699. All of these logs are called logarithms to base 10 because they are multiples of 10. Expressed as $\log_{10}$, whole numbers and fractions can be looked up in tables, determined by a slide rule, or produced by a calculator. Use a calculator!

Converting large numbers to their $\log_{10}$ value illustrates that for huge numbers of microbes, doubling, tripling, and even quadrupling does not mean much in microbiology. The log of $1.18 \times 10^7$ equals 7.07. Doubling $1.18 \times 10^7$ to $2.36 \times 10^7$ results in a $\log_{10}$ of 7.37, not 14.14 (2 times 7.07). The triple of $1.18 \times 10^7$ is $3.54 \times 10^7$ or $\log_{10}$ equal to 7.55; quadrupling the number gives a $\log_{10}$ of 7.67. This illustrates that bacterial numbers differing by a few multiples can be viewed as being of the same general magnitude. Only when bacterial numbers change by at least 100 times do microbiologists view this as a real change beyond the normal variability of nature.

## Anaerobic microbiology

Diluting and counting anaerobic bacteria resembles the steps used for aerobic bacteria except that anaerobes require sealed containers that exclude all air. Anaerobic microbiology calls for diligence that aerobic methods ignore, that is, the microbiologist follow aseptic techniques *and* keep air away from the bacteria.

Anaerobic bacteria grow only on agar plates placed inside a sealed jar containing a chemical to remove all the oxygen from the jar once it has been sealed. As a second option, microbiologists can use an anaerobic chamber, which is a large plastic bubble filled with an inert gas lacking oxygen. One side of the chamber has arm holes built directly into the plastic so that a microbiologist can sit outside the chamber, put her arms into the arm holes, and dilute and perform other activities with the anaerobes inside the chamber. Some anaerobic chambers include a small incubator so that plates need never exit the anaerobic environment during an experiment.

I learned anaerobic microbiology by using a third method named for Robert Hungate who advanced the techniques for growing strict anaerobes in the 1950s and 1960s. The Hungate method developed almost exclusively by studying the anaerobes from the digestive tracts of cattle, sheep, and goats. These bacteria have more stringent requirements for oxygen-free environments—they are often referred to as fastidious anaerobes. The Hungate method thus grows the bacteria in test tubes instead of plates, which are impractical for airtight conditions.

Hungate tubes are prepared by pouring sterile molten agar into each tube and then inoculating the agar while it is still a liquid. Microbiologists exclude air from the open tube during this step by directing a gentle stream of inert gas into the tube. The microbiologist must inoculate the agar quickly and then withdraw the gassing hose an instant before sealing the tube with a rubber stopper. Fastidious anaerobes require stoppers made of special rubber that prevents any molecule of air from seeping into the tube during incubation. A good practitioner of anaerobic microbiology can perform the one-two step of withdrawing the hose and stoppering the tube quicker than the eye can follow. The microbiologist then rolls the inoculated tubes on a horizontal surface until the agar has solidified into a uniform layer coating the inside of the tube. After incubation, the microbiologist counts CFUs in the agar.

## Aseptic technique

All microbiological procedures require aseptic technique, which refers to all the activities microbiologists perform to keep unwanted microbes out of pure cultures or sterile items. Aseptic means free from germs, and sepsis is a medical term for the presence of germs.

Media, glassware, and anything else that comes in contact with live cultures must be sterilized in an autoclave. This piece of equipment treats liquids and solids with pressurized steam to kill all microbes. Items that have been sterilized and covered can be stored indefinitely.

In addition to sterilized laboratory supplies, microbiologists also "flame" items over a Bunsen burner before handling bacterial cultures. Flaming works well for metal or glass items such as inoculating loops, forceps, and open test tubes.

All these activities require that the microbiologist imagine where bacteria exist and predict the places most likely to suffer contamination. To reduce the chances of contamination by unseen and unwanted microbes, aseptic technique includes disinfection of laboratory surfaces before and after using them. Microbiologists also avoid coughing, sneezing, and breathing into open culture containers.

Surgery rooms exemplify aseptic technique because every action performed there is done in a manner to prevent contamination of the patient. Aseptic technique does not require sophisticated technology, but neither does it tolerate shortcuts. Whatever scientific advances microbiology absorbs in the future, aseptic techniques will endure in much the same way they are practiced today.

# Resources for learning more about bacteria

## Internet resources on bacteria

Bacteria World: http://www.bacteria-world.com/.

Cells Alive: http://www.cellsalive.com/.

Dennis Kunkel Microscopy: http://www.denniskunkel.com/.

Infectious Diseases in History: http://urbanrim.org.uk/diseases.htm.

Microbe World: http://www.microbeworld.org/.

Todar's Online Textbook of Bacteriology: http://www.textbookofbacteriology.net/.

The Microbial World: http://www.microbiologytext.com/index.php?module=Book&func=toc&book_id=4.

University of California Museum of Paleontology: http://www.ucmp.berkeley.edu/bacteria/bacteria.html.

The Virtual Museum of Bacteria: http://www.bacteriamuseum.org/cms/.

## Book resources on bacteria

Biddle, Wayne. *A Field Guide to Germs*, 2002, Anchor Books, New York.

Dyer, Betsey Dexter. *A Field Guide to the Bacteria*, 2003, Cornell University Press, Ithaca, NY.

Lax, Alistair. *Toxin: The Cunning of Bacterial Poisons*, 2005, Oxford University Press, Oxford.

Maczulak, Anne E. *The Five-Second Rule and Other Myths about Germs*, 2007, Thunder's Mouth Press/Perseus Books, Philadelphia.

Meinesz, Alexandre. *How Life Began, Evolution's Three Geneses*, 2008, University of Chicago Press.

Sachs, Jessica Snyder. *Good Germs, Bad Germs: Health and Survival in a Bacterial World*, 2007, Hill and Wang, New York.

Schaechter, Moselio, John L. Ingraham, and Frederick C. Neidhardt. *Microbe*, 2006, American Society for Microbiology Press, Washington, DC.

Spellberg, Brad. *Rising Plague: The Global Threat from Deadly Bacteria and Our Dwindling Arsenal to Fight Them*, 2009, Prometheus Books, New York.

Zimmer, Carl. *Microcosm: E. coli and the New Science of Life*, 2008, Vintage Books, New York.

## Classic reading on bacteria

De Kruif, Paul. *Microbe Hunters*, 1926, Harcourt, Orlando, Fla. History of bacteriology through biographies of the greatest microbiologists.

Garrett, Laurie. *The Coming Plague: Newly Emerging Diseases in a World Out of Balance*, 1994, Farrar, Straus and Giroux, New York. Mainly about viruses but with eternal lessons on all germs.

Karlen, Arlo. *Biography of a Germ*, 2000, Pantheon Books, New York. A unique introduction to bacteria by following the activities of the lyme disease pathogen, *Borrelia burgdorferi*.

MacFarlane, Gwyn. *Alexander Fleming: The Man and the Myth*, 1985, Oxford University Press, Oxford. The story and intrigue behind a historic scientific discovery.

Thomas, Lewis. *The Lives of a Cell: Notes of a Biology Watcher*, 1974, Viking Press, New York. Appreciation of biology for nonbiologists.

# Bacteria rule references

## Chapter 1

### *Print*

Brothwell, D. R., and P. Brothwell. *Food in Antiquity*. Johns Hopkins University Press, 1998.

Luckey, Thomas D. "Effects of Microbes on Germfree Animals." In *Advances in Applied Microbiology*, Volume 7. Edited by Wayne W. Umbreit. Academic Press, 1965.

Munn, Colin B. *Marine Microbiology—Ecology and Applications*. New York: Garland Science, 2004.

Rainey, Fred A., and Aheron Oren, eds. *Extremophiles*. London: Elsevier, 2006.

### *Internet*

Astrobiology Web. "Life in Extreme Environments." http://www.astrobiology.com/extreme.html#archaea.

Baron, Samuel, ed. "Normal Flora of Skin. Chap. 6 in *Medical Microbiology*, 4th ed. Galveston: University of Texas Medical Branch, 1996. http://www.ncbi.nlm.nih.gov/bookshelf/br.fcgi?book=mmed&part=A512.

BBC News. "A Whale of a Bug." News release, April 15, 1999. http://news.bbc.co.uk/2/hi/science/nature/320117.stm.

Central Vermont Public Service. http://www.cvps.com.

Chung, King-Thom, and Christine L. Case. "Sergei Winogradsky: Founder of Soil Microbiology." *Society for Industrial Microbiology News* 51 (2001): 133-35. http://www.smccd.edu/accounts/case/envmic/winogradsky.pdf.

CNN.com. "Star Survey Reaches 70 Sextillion." News release, July 23, 2003. http://www.cnn.com/2003/TECH/space/07/22/stars.survey.

DeLong, Edward F. "Archaeal Mysteries of the Deep Revealed. *Proceedings of the National Academy of Sciences* 103 (2006): 6417-18. http://www.ncbi.nlm.nih.gov/pmc/articles/PMC1458900.

DuBois, Andre. "Spiral Bacteria in the Human Stomach: The Gastric Helicobacters." *Emerging Infectious Diseases* 1 (1995): 79-88. http://www.cdc.gov/ncidod/eid/vol1no3/dubois.htm.

Eder, Waltraud, and Erika von Mutius. "Hygiene Hypothesis and Endotoxin: What is the Evidence?" *Current Opinions in Clinical Immunology* 4 (2004): 113-17. http://www.forallvent.info/uploads/media/von_Mutius_Hygiene_hypothesis_and_endotoxin_2004.pdf.

Egland, Paul G., Robert J. Palmer, and Paul E. Kolenbrander. "Interspecies Communication in *Streptococcus gordonii—Veillonella atypica* Biofilms: Signaling in Flow Conditions Requires Juxtaposition." *Proceedings of the National Academy of Sciences* 101 (2004): 16917-22. http://www.pnas.org/content/101/48/16917.full.

*Eureka Science News*. "A Woman's Nose Knows Body Odor." April 6, 2009. http://esciencenews.com/articles/2009/04/07/a.womans.nose.knows.body.odor.

Favier, Christine F., Willem M. de Vos, Antoon D. L. Akkermans. "Development of Bacterial and Bifidobacterial Communities in Feces of Newborn Babies." *Anaerobe* 9 (2003): 219-29. http://www.sciencedirect.com/science?_ob=ArticleURL&_udi=B6W9T-49J8TBH-1&_user=10&_rdoc=1&_fmt=&_orig=search&_sort=d&_docanchor=&view=c&_searchStrId=1024766517&_rerunOrigin=google&_acct=C000050221&_version=1&_urlVersion=0&_userid=10&md5=c3bab74e038cbd08afbd0e897c31ee96.

Featherstone, J. D. B. "The Continuum of Dental Caries—Evidence for a Dynamic Disease Process." *Journal of Dental Research* 83 (2004): C39-C42. http://jdr.sagepub.com/cgi/reprint/83/suppl_1/C39.pdf.

Folk, Robert L. "Nanobacteria: Surely Not Figments, but under What Heaven Are They?" *Natural Science*, February 11, 1997. http:/ /naturalscience.com/ns/articles/01-03/ns_folk.html.

Food and Agricultural Organization of the United Nations. "Hydrogen Production." Chap. 5 in *Renewable Biological Systems for Alternative Sustainable Energy Production*. Edited by Kasuhisha Miyamoto. FOA, 1997. http://www.fao.org/docrep/w7241e/w7241e0g.htm#5.2%20biophotolysis%20of%20water%20by%20microalgae%20and%20cyanobacteria.

Fulhage, Charles D., Dennis Sievers, and James R. Fischer. "Generating Methane Gas from Manure." University of Missouri Extension, October 1993. http://extension.missouri.edu/publications/DisplayPub.aspx?P=G1881.

Handwerk, Brian. "Armpits Are 'Rain Forests' for Bacteria, Skin Map Shows." *National Geographic News*, May 28, 2009. http://news.nationalgeographic.com/news/2009/05/090528-armpits-bacteria-rainforests.html.

Helicobacter Foundation. 2006. http://www.helico.com/h_general.html.

Higaki, Shuichi, Taro Kitagawa, Masaaki Morohashi, and Takayoshi Tamagishi. "Anerobes Isolated From Infectious Skin Diseases." *Anaerobe* 5 (1999): 583-87. http://www.sciencedirect.com/science?_ob=ArticleURL&_udi=B6W9T-45HR7PX-C&_user=10&_rdoc=1&_fmt=&_orig=search&_sort=d&_docanchor=&view=c&_searchStrId=1021747765&_rerunOrigin=google&_acct=C000050221&_version=1&_urlVersion=0&_userid=10&md5=ee63538d7839935b73ce52130111d3ae.

Keith, William A., Roko J. Smiljanec, William A. Akers, ad Lonnie W. Keith. "Uneven Distribution of Aerobic Mesophilic Bacteria on Human Skin." *Applied and Environmental Microbiology* 37 (1979): 345-47. http://aem.asm.org/cgi/reprint/37/2/345.pdf.

Krulwich, Robert. "Bacteria Outnumber Cells in Human Body." National Public Radio. *All Things Considered*, July 1, 2006. http://www.npr.org/templates/story/story.php?storyId=5527426.

Lubick, Naomi. "Where Biosphere Meets Geosphere." *Scientific American*, January 28, 2002. http://www.scientificamerican.com/article.cfm?id=where-biosphere-meets-geo.

Martinez, Chelsea. "Baby's First Bacteria." *Los Angeles Times*, June 26, 2007. http://www.latimes.com/features/health/la-hew-booster26jun26,1,3571421.story.

Montville, Thomas J. "Dependence of *Clostridium botulinum* Gas and Protease Production on Culture Conditions." *Applied and Environmental Microbiology* 45 (1983): 571-75. http://www.ncbi.nlm.nih.gov/pmc/articles/PMC242325/pdf/aem00171-0229.pdf.

National Human Genome Research Institute. http://www.genome.gov/10000354#top.

National Science Foundation. "Bacteria May Thrive in Antarctic Lake." News release, December 9, 1999. http://www.nsf.gov/od/lpa/news/press/99/pr9972.htm.

Nobel Foundation. "Marshall, Barry J., and J. Robin Warren." Nobel Prize Press release, October 3, 2005. http://nobelprize.org/nobel_prizes/medicine/laureates/2005/press.html.

Pasteur, Louis. *Oeuvres de Pasteur*. Liebraires de l'Académie de Médicine, Paris, 1922.

Peterson, W. H., and Mary S. Peterson. "Relation of Bacteria to Vitamins and Other Growth Factors." *Bacteriological Reviews* 9 (1946): 49-109. http://www.ncbi.nlm.nih.gov/pmc/articles/PMC440891.

Porter, Alan M. W. "Why Do we Have Apocrine and Sebaceous Glands." *Journal of the Royal Society of Medicine* 94 (2001): 236-37. http://jrsm.rsmjournals.com/cgi/content/full/94/5/236.

Sapp, Jan. "The Prokaryote-Eukaryote Dichotomy: Meanings and Mythology." *Microbiology and Molecular Biology Reviews* 69 (2005): 292-305. http://mmbr.asm.org/cgi/content/full/69/2/292.

*ScienceDaily*. "Human Skin Harbors Completely Unknown Bacteria." February 6, 2007. http://www.sciencedaily.com/releases/2007/02/070206095816.htm.

————. "The Hygiene Hypothesis: Are Cleanlier Lifestyles Causing More Allergies for Kids?" September 9, 2007. http://www.sciencedaily.com/releases/2007/09/070905174501.htm.

————. "A Survivor in Greenland: A Novel Bacterial Species Is Found Trapped in 120,000-Year-old Ice." June 3, 2008. http://www.sciencedaily.com/releases/2008/06/080603104418.htm.

Stark, P. L., and A. Lee. "The Bacterial Colonization of the Large Bowel of Pre-term Low Birth Weight Neonates." *Journal of Hygiene* 89 (1982): 59-67. http://www.ncbi.nlm.nih.gov/pubmed/7097003.

University of Georgia. "First-ever Estimate of Total Bacteria on Earth." *San Diego Earth Times*, September 1988. http://www.sdearthtimes.com/et0998/et0998s8.html.

University of Iowa. "Turning on Cell-Cell Communication Wipes Out Staph Biofilms." News release, April 30, 2008. http://www.news-releases.uiowa.edu/2008/april/043008biofilms.html.

Vaglio, Stefano. "Chemical Communication and Mother-Infant Recognition." *Communicative and Integrative Biology* 2 (2009): 279-81. http://www.ncbi.nlm.nih.gov/pmc/articles/PMC2717541.

Von Mutius, Erika. "A Conundrum of Modern Times That's Still Unresolved." *European Respiratory Journal* 22 (2003): 719-720. http://www.erj.ersjournals.com/cgi/reprint/22/5/719.

Wassenaar, T. M. "Extremophiles." Virtual Museum of Bacteria. January 6, 2009. http://www.bacteriamuseum.org/cms/Evolution/extremophiles.html.

## Chapter 2

### *Print*

Cano, Raul J., and Monica K. Borucki. "Revival and Identification of Bacterial Spores in 25- to 40-Million-year-old Dominican Amber." *Science* 268 (1995): 1060-64.

Cantor, Norman F. *In the Wake of the Plague*. New York: Simon and Schuster, 2001.

Debré, Patrice. *Louis Pasteur*. Translated by Elborg Forster. Johns Hopkins University Press, 1994.

Fleming, Alexander. "On the Antibacterial Action of Cultures of a *Penicillium*, with Special Reference to their Use in the Isolation of *B. influenzae*." *British Journal of Experimental Pathology* 10 (1929): 226-36.

Fribourg-Blanc, A., and H. H. Mollaret. "Natural Treponematosis of the African Primate." *Primate Medicine* 3 (1969): 113-21.

Fribourg-Blanc, A., H. H. Mollaret, and G. Niel. "Serologic and Microscopic Confirmation of Treponematosis in Guinea Baboons." *Bulletin of the Exotic Pathology Society* 59 (1966): 54-59.

Garrison, Fielding H. *An Introduction to the History of Medicine*. W. B. Saunders & Co., 1921.

Geison, Gerald L. *The Private Science of Louis Pasteur*. Princeton University Press, 1995.

Horan, Julie L., and Deborah Frazier. *The Porcelain God*. Carol Publishing Corp., 1996.

Leavitt, Judith Walzer. *Typhoid Mary: Captive to the Public's Health*. Beacon Press, 1996.

Leon, Ernestine F. "A Case of Tuberculosis in the Roman Aristocracy at the Beginning of the Second Century." *Journal of the History of Medicine and Allied Sciences* 64 (1959): 86-88.

Livi-Bacci, Massimo. *A Concise History of World Population*, 3rd ed. Blackwell, 2001.

Zivanovic, Srboljub, and L. F. Edwards. *Ancient Diseases: Elements of Paleopathology*. Methuen and Co., 1982.

### Internet

American University. "The Role of Trade in Transmitting the Black Death." http://www1.american.edu/TED/bubonic.htm.

Atlas, R. M. "*Legionella*: From Environmental Habitats to Disease Pathology, Infection and Control." *Environmental Microbiology* 1 (1999): 283-93. http://www.ncbi.nlm.nih.gov/pubmed/11207747.

Barbaree, J. M., B. S. Fields, J. C. Feeley, G. W. Gorman, and W. T. Martin. "Isolation of Protozoa from Water Associated with a Legionellosis Outbreak and Demonstration of Intracellular Multiplication of *Legionella pneumophila*." *Applied and Environmental Microbiology* 51 (1986): 422-24. http://aem.asm.org/cgi/content/abstract/51/2/422.

BBC News. "Alive...after 250 Million Years." October 18, 2000. http://news.bbc.co.uk/2/hi/science/nature/978774.stm.

———. "Legionnaires' Disease—A History of Its Discovery." January 16, 2003. http://www.bbc.co.uk/dna/h2g2/A882371.

Bidle, Kay D., SangHoon Lee, David R. Marchant, and Paul G. Falkowski. "Fossil Genes and Microbes in the Oldest Ice on Earth." *Proceedings of the National Academy of Sciences* 104 (2007):13455-60. http://www.pnas.org/content/104/33/13455.short.

Brown, Michael. "Ancient DNA, Found Mostly in Amber-preserved Specimens." Molecular History Research Center. http://www.mhrc.net/ancientDNA.htm.

Calloway, Ewen. "Ancient Bones Show Earliest 'Human' Infection." New Scientist, August 2009. http://www.newscientist.com/article/dn17559-ancient-bones-show-earliest-human-infection.html.

Evans, James Allan. "Justinian (527-565 A.D.)." http://www.roman-emperors.org/justinia.htm.

Gallagher, Patricia E., and Stephen J. Greenberg. "The History of Diseases." 2009. http://www.mla-hhss.org/histdis.htm.

Gilbert, Geoffrey. *World Population*, 2nd ed. Santa Barbara, Calif.: ABC-CLIO, 2005.

Harvard University. Contagion: Historical Views of Diseases and Epidemics. http://ocp.hul.harvard.edu/contagion.

Hippocrates. *Aphorisms*. http://classics.mit.edu/Hippocrates/aphorisms.5.v.html.

Iliffe, Rob. "Robert Hooke's Critique of Newton's Theory of Light and Colors (Delivered 1672)." The Newton Project. http://www.newtonproject.sussex.ac.uk/view/texts/normalized/NATP00005.

Internet Modern History Sourcebook. "Louis Pasteur (1822-1895): Germ Theory and Its Applications to Medicine and Surgery, 1878." http://www.fordham.edu/halsall/mod/1878pasteur-germ.html.

Kilpatrick, Howard J., and Andrew M. LaBonte.. *Managing Urban Deer in Connecticut*, 2nd ed. Connecticut Department of Environmental Protection Bureau of Natural Resources, 2007. http://www.ct.gov/dph/lib/dph/urbandeer07.pdf.

Kwaik, Yousef Abu, Lian-Yong Gao, Barbara J. Stone, Chandrasekar Venkataraman, and Omar S. Harb. "Invasion of Protozoa by *Legionella pneumophila* and Its Role in Bacterial Ecology and Pathogenesis." *Applied and Environmental Microbiology* 9 (1998): 3127-33. http://aem.asm.org/cgi/content/full/64/9/3127.

Lane, Samuel. "A Course of Lectures on Syphilis." *Lancet* 1 (1841): 217-23. http://books.google.com/books?id=gfsBAAAAYAAJ&pg=PA219&dq=seige+of+naples+syphilis#v=onepage&q=&f=false.

Lanoil, Brian, Mark Skidmore, John C. Priscu, Sukkyun Han, Wilson Foo, Stefan W. Vogel, Slawek Tulaczyk, and Hermann Engelhardt. "Bacteria Beneath the West Antarctic Ice Sheet." *Environmental Microbiology* 11 (2009): 609-15. http://www.homepage.montana.edu/~lkbonney/DOCS/Publications/LanoilEtAlBacteriaWAIS.pdf.

Loghem, J. J. van. "The Classification of the Plague-Bacillus." *Antonie van Leeuwenhoek* 10 (1944): 15-16. http://www.springerlink.com/content/k65773478g138348.

Madigan, Michael T., and Barry L. Marrs. "Extremophiles." *Scientific American*, April 1997. http://atropos.as.arizona.edu/aiz/teaching/a204/extremophile.pdf.

Maugh, Thomas H. "An Empire's Epidemic." *Los Angeles Times*, May 6, 2002. http://www.ph.ucla.edu/EPI/bioter/anempiresepidemic.html.

Medical Front WW I. http://www.vlib.us/medical/Nindex.htm.

Microbe World. "Oldest Living Microbes." January 14, 2009. http://www.microbeworld.org/index.php?option=com_content&view=article&id=156&Itemid=87.

*Micscape*. "Robert Hooke." March 2000. http://www.microscopy-uk.org.uk/mag/indexmag.html?http://www.microscopy-uk.org.uk/mag/artmar00/hooke1.html.

Molmeret, Maëlle, Matthias Horn, Michael Wagner, Marina Santic, and Yousef Abu Kwaik. "Amoebae as Training Grounds for Intracellular Bacterial Pathogens." *Applied and Environmental Microbiology* 71 (2005): 20-28. http://aem.asm.org/cgi/content/full/71/1/20.

Mulder, Henry. "Newton and Hooke: A Tale of Two Giants." Science and You, 2008. http://www.scienceandyou.org/articles/ess_14.shtml.

*New World Encyclopedia*. "Hooke, Robert." http://www.newworldencyclopedia.org/entry/Robert_Hooke.

NJMS National Tuberculosis Center. "Brief History of Tuberculosis." http://www.umdnj.edu/~ntbcweb/history.htm.

Nummer, Brian A. "Historical Origins of Food Preservation." National Center for Home Food Preservation, May 2002. http://www.uga.edu/nchfp/publications/nchfp/factsheets/food_pres_hist.html.

O'Connor, Anahad. "Dr. Norman Heatley, 92, Dies; Pioneer in Penicillin Supply." *The New York Times*, January 17, 2004. http://www.nytimes.com/2004/01/17/world/dr-norman-heatley-92-dies-pioneer-in-penicillin-supply.html.

PBS. "Deciphering Disease in Ancient Mummies." http://www.pbs.org/wnet/pharaohs/secrets4.html.

Raleigh, Veena Soni. "Trends in World Population: How Will the Millennium Compare with the Past?" *Human Reproduction Update* 5 (1999): 500-05. http://humupd.oxfordjournals.org/cgi/reprint/5/5/500.pdf.

RobertHooke.com. http://www.roberthooke.com/Default.htm.

Rosner, David. "Beyond Typhoid Mary: The Origins of Public Health at Columbia and in the City." *Columbia Magazine*, Spring 2004. http://www.columbia.edu/cu/alumni/Magazine/Spring2004/publichealth.html.

Shulman, Matthew. "12 Diseases that Altered History." *U.S. News and World Report*, January 3, 2008. http://www.usnews.com/health/articles/2008/01/03/12-diseases-that-altered-history.html.

Travis, J. "Prehistoric Bacteria Revived from Buried Salt." *Science News* 155 (1999): 398. http://www.sciencenews.org/sn_arc99/6_12_99/fob3.htm.

Tschanz, David W. "Typhus Fever on the Eastern Front in World War I." http://entomology.montana.edu/historybug/WWI/TEF.htm.

University of Pittsburgh. Supercourse: Cholera—History. http://www.pitt.edu/~super1/lecture/lec1151/index.htm.

Vreeland, Russell H., William D. Rosenzweig, and Dennis W. Powers. "Isolation of a 250-Million-year-old Halotolerant Bacterium from a Primary Salt Crystal." *Nature* 407 (2000): 897-900. http://www.nature.com/nature/journal/v407/n6806/full/407897a0.html#B2.

Waggoner, Ben. "Robert Hooke." http://www.ucmp.berkeley.edu/history/hooke.html.

## Chapter 3

### *Print*

Abraham, E. P., and E. Chain. "An Enzyme from Bacteria Able to Destroy Penicillin." *Nature* 3713 (1940): 836. In *Microbiology: A Centenary Perspective*. Edited by Wolfgang K. Jolik, Lars G. Ljungdahl, Alison D. O'Brien, Alexander von Graevenitz, and Charles Yanofsky. American Society of Microbiology Press, 1999.

Chain, E., H. W. Florey, A. D. Gardner, N. G. Heatley, M. A. Jennings, J. Orr-Ewing, and A. G. Sanders. "Penicillin as a Chemotherapeutic Agent." *Lancet* 2 (1940): 226-28. In *Microbiology: A Centenary Perspective*. Edited by Wolfgang K. Jolik, Lars G. Ljungdahl, Alison D. O'Brien, Alexander von Graevenitz, and Charles Yanofsky. American Society of Microbiology Press, 1999.

Levy, Stuart B. *The Antibiotic Paradox: How the Misuse of Antibiotics Destroys Their Curative Powers*. Perseus Publishing, 2002.

Mayhall, G. Glen. *Hospital Epidemiology and Infection Control*, 3rd ed. Lippincott Williams and Wilkins, 2004.

Ponte-Sucre, Alicia, ed. *ABC Transporters in Microorganisms*. Caister Academic Press, 2009.

### Internet

Armstrong, J. L., D. S. Shigeno, J. J. Calomiris, and R. J. Seidler. "Antibiotic-resistant Bacteria in Drinking Water." *Applied and Environmental Microbiology* 42 (1981): 277-83. http://www.ncbi.nlm.nih.gov/pmc/articles/PMC244002.

Australian Institute of Marine Science. "New Marine Antibiotics to Fight Disease." News release, May 13, 2004. http://www.aims.gov.au/news/pages/media-release-20040513.html.

Braibant, M., P. Gilot, and J. Content. "The ATP Binding Cassette (ABC) Transport Systems of *Mycobacterium tuberculosis*." *FEMS Microbiological Reviews* 24 (2000): 449-67. http://www.ncbi.nlm.nih.gov/pubmed/10978546.

Bryner, Jeanna. "Fight Against Germs May Fuel Allergy Increase." Fox News, September 17, 2007. http://www.foxnews.com/story/0,2933,296869,00.html?sPage=fnc/scitech/humanbody.

Choi, Charles Q. "Antibiotic-Resistance DNA Showing Up in Drinking Water." Fox News, November 2, 2006. http://www.foxnews.com/story/0,2933,227106,00.html.

Choi, Cheol-Hee. "ABC Transporters as Multidrug Resistance Mechanisms and the Development of Chemosensitizers for Their Reversal." *Cancer Cell International* 5 (2005): 30-43. http://www.cancerci.com/content/pdf/1475-2867-5-30.pdf.

Dougherty, Elizabeth. "Bacterial Viruses Boost Antibiotic Action." *Harvard Focus*, April 3, 2009. http://focus.hms.harvard.edu/2009/040309/biomedical_engineering.shtml.

Falda, Wayne. "N. D. Prof Part of 'Trojan Horse' Discovery." *South Bend Tribune*, September 29, 2000. http://www.mcgill.ca/files/microimm/coulton_article_southbendtribune.pdf.

Fleming, Alexander. "Penicillin." Nobel lecture presented December 11, 1945. http://nobelprize.org/nobel_prizes/medicine/laureates/1945/fleming-lecture.pdf.

Florey, Howard W. "Penicillin." Nobel lecture presented December 11, 1945. http://nobelprize.org/nobel_prizes/medicine/laureates/1945/florey-lecture.pdf.

Harrell, Eben. "The Desperate Need for New Antibiotics." *Time*, October 1, 2009. http://www.time.com/time/health/article/0,8599,1926853,00.html.

Isnansetyo, Alim, and Yuto Kamei. "MC-21A, a Bactericidal Antibiotic Produced by a New Bacterium, *Pseudoalteromonas phenolica* sp. nov. O-BC30$^{T}$, against Methícillin-resistant *Staphylococcus aureus*." *Antimicrobial Agents and Chemotherapy* 47 (2003): 480-88. http://www.ncbi.nlm.nih.gov/pmc/articles/PMC151744.

Johansen, Helle Krogh, Thøger Gorm Jensen, Ram Benny Dessau, Bettina Lundgren, and Niels Fremodt-Møller. "Antagonism between penicillin and erythromycin against *Streptococcus pneumoniae in vitro* and *in vivo*." *Journal of Antimicrobial Chemotherapy* 46 (2000): 973-80. http://jac.oxfordjournals.org/cgi/reprint/46/6/973.

Karthikeyan, K. G., and Michael T. Meyer. "Occurrence of Antibiotics in Wastewater Treatment Facilities in Wisconsin, USA." *Science of the Total Environment* 361 (2006): 196-207. http://www.sciencedirect.com/science?_ob=ArticleURL&_udi=B6V78-4GTW8Y0-2&_user=10&_rdoc=1&_fmt=&_orig=search&_sort=d&_docanchor=&view=c&_searchStrId=1135533811&_rerunOrigin=google&_acct=C000050221&_version=1&_urlVersion=0&_userid=10&md5=492067b7c9f9fb6934246de18ab05811.

Kerr, I. D., E. D. Reynolds, and J. H. Cove. "ABC Proteins and Antibiotic Drug Resistance: Is It All About Transport?" *Biochemical Society Transactions* 33 (2005): 1000-1002. http://www.biochemsoctrans.org/bst/033/1000/0331000.pdf.

Kümmerer, K. "Resistance in the Environment." *Journal of Antimicrobial Chemotherapy* 54 (2004): 311-20. http://jac.oxfordjournals.org/cgi/reprint/54/2/311.

McKibben, Linda, Teresa Horan, Jerome I. Tokars, Gabrielle Fowler, Denise M. Cardo, Michele L. Pearson, and Patrick J. Brennan. "Guidance on Public Reporting of Healthcare-associated Infections: Recommendations of the Healthcare Infection Control Practices Advisory Committee." *American Journal of Infection Control* 33 (2005): 217-26. http://www.cdc.gov/ncidod/dhqp/pdf/hicpac/PublicReportingGuide.pdf.

National Oceanic and Atmospheric Administration. "Antibiotic Resistance: A Rising Concern in Marine Ecosystems." February 13, 2009. http://www.noaanews.noaa.gov/stories2009/20090213_antibiotic.html.

PBS. "Antibiotic Debate Overview." http://www.pbs.org/wgbh/pages/frontline/shows/meat/safe/overview.html.

PhysOrg.com. "Toward Improved Antibiotics Using Proteins from Marine Diatoms." September 8, 2008. http://www.physorg.com/news140112686.html.

Reinthaler, F. F., J. Posch, G. Feierl, G. Wüst, D. Haas, G. Ruckenbauer, F. Mascher, and E. Marth. "Antibiotic Resistance of *E. coli* in Sewage and Sludge." *Water Research* 37 (2003): 1685-90. http://www.ncbi.nlm.nih.gov/pubmed/12697213.

*ScienceDaily*. "Trojan Horse Strategy Defeats Drug-resistant Bacteria." March 17, 2007. http://www.sciencedaily.com/releases/2007/03/070316091659.htm.

Siegel, Jane D., Emily Rhinehart, Marguerite Jackson, and Linda Chiarello. "Management of Multidrug-resistant Organisms in Healthcare Settings, 2006." Centers for Disease Control and Prevention. http://www.cdc.gov/ncidod/dhqp/pdf/ar/mdroguideline2006.pdf.

Sociology, History and Philosophy Resource Center. "Penicillin and Chance." http://www1.umn.edu/ships/updates/fleming.htm.

Soga, Yoshihigo, Takashi Saito, Fusanori Ishimaru, Junji Mineshiba, Fumi Mineshiba, Hirokazu Takaya, Hideaki Sato, et al. "Appearance of Multidrug-resistant Opportunistic Bacteria on the Gingiva during Leukemia Treatment." *Journal of Periodontology* 79 (2008): 181-86. http://www.joponline.org/doi/abs/10.1902/jop.2008.070205%20?url_ver=Z39.88-2003&rfr_id=ori:rid:crossref.org&rfr_dat=cr_pub%3Dncbi.nlm.nih.gov.

Watkinson, A. J., G. B. Micalizzi, G. M. Graham, J. B. Bates, and S. D. Costanzo. "Antibiotic-resistant *Escherichia coli* in Wastewaters, Surface Waters, and Oysters from an Urban Riverine System." *Applied and Environmental Microbiology* 73 (2007): 5667-70. http://www.ncbi.nlm.nih.gov/pmc/articles/PMC2042091.

Wilson, Richard. "Penicillin Overuse Puts Fleming's Legacy at Risk." *London Sunday Times*, September 7, 2008. http://www.timesonline.co.uk/tol/news/uk/scotland/article4691534.ece.

## Chapter 4

### *Print*

Antonioli, P., G. Zapparoli, P. Abbruscato, C. Sorlini, G. Ranalli, and P.G. Righetti. "Art-loving Bugs: The Resurrection of Spinello Aretino from Pisa's Cemetery." *Proteomics* 5 (2005): 2453-59.

Crichton, Michael. *The Andromeda Strain*. New York: Alfred A. Knopf, 1987.

Mann, Thomas. *Death in Venice*. 1912.

Maugham, W. Somerset. *The Painted Veil*, 1925.

Rao, T. S., S. N. Sairam, B. Viswanathan, and K. V. K. Nair. "Carbon Steel Corrosion by Iron Oxidizing and Sulphate Reducing Bacteria in a Freshwater Cooling System. *Corrosion Science* 42 (2000): 1417-31.

### *Internet*

Arenskötter, M., D. Baumeister, M. M. Berekaa, G. Pötter, R. M. Kroppenstedt, A. Linos, and A. Steinbüchel. "Taxonomic Characterization of Two Rubber Degrading Bacteria Belonging to the Species *Gordonia polyisoprevivorans* and Analysis of Hyper Variable Regions of 16S rDNA Sequences." *FEMS Microbiology Letters* 205 (2001): 277-81. http://www3.interscience.wiley.com/cgi-bin/fulltext/119020815/PDFSTART?CRETRY=1&SRETRY=0.

Bröker, D., D. Dietz, M. Arenskötter, and A. Steinbüchel. "The Genomes of the Non-clearing-zone-forming Species *Gordonia polyisoprevivorans* and *Gordonia westfalica* Harbor Genes Expressing Lcp Activity in *Streptomyces* Strains." *Applied and Environmental Microbiology* 74 (2008): 2288-97. http://www.ncbi.nlm.nih.gov/pubmed/18296529.

Cappitelli, Francesca, Lucia Toniolo, Antonio Sansonetti, Davide Gulotta, Giancarlo Ranalli, Elisabetta Zanardini, and Claudia Sorlini. "Advantages of Using Microbial Technology over Traditional Chemical Technology in Removal of Black Crusts from Stone Surfaces of Historical Monuments." *Applied and Environmental Microbiology* 73 (2007): 5671-75. http://aem.asm.org/cgi/reprint/73/17/5671.pdf.

Chalke, H. D. "The Impact of Tuberculosis on History, Literature and Art." *Medical History* 6 (1962): 301-18. http://www.ncbi.nlm.nih.gov/pmc/articles/PMC1034755.

Ciferri, Orio. "Microbial Degradation of Paintings." *Applied and Environmental Microbiology* 65 (1999): 879-85. http://www.ncbi.nlm.nih.gov/pmc/articles/PMC91117.

Dardes, Kathleen, and Andrea Rothe, eds. *The Structural Conservation of Panel Paintings*. The Getty Conservation Institute, 1995. http://www.getty.edu/conservation/publications/pdf_publications/panelpaintings1.pdf.

*Eureka Science News*. "Biotech Scientists Team with Curators to Stem Decay of World's Art, Cultural Heritage." February 8, 2009. http://esciencenews.com/articles/2009/02/08/biotech.scientists.team.with.curators.stem.decay.worlds.art.cultural.heritage.

Gupta, M., and D. Alcid. "A Rubber-degrading Organism Growing from a Human Body." *International Journal of Infectious Diseases* 12 (2008): e332-e333. http://www.ijidonline.com/article/S1201-9712(08)01020-5/abstract.

Gupta, M., D. Prasad, H. S. Khara, and D. Alcid. "A Rubber-degrading Organism Growing from a Human Body." *Journal of Infectious Diseases* 14 (2010): e75-e76. http://www.ncbi.nlm.nih.gov/pubmed/19501006.

Harmon, Katherine. "The Science of Saving Art: Can Microbes Protect Masterpieces?" *Scientific American*, February 9, 2009.

Jendrossek, D., G. Tomasi, and R. M. Kroppenstedt. "Bacterial Degradation of Natural Rubber: A Privilege of Actinomycetes?" *FEMS Microbiology Letters* 150 (1997): 179-88. http://grande.nal.usda.gov/ibids/index.php?mode2=detail&origin=ibids_references&therow=138921.

Kerksiek, Kristen. "The Art of Infection." *Infection Research*, October 29, 2009. http://www.infection-research.de/perspectives/detail/pressrelease/the_art_of_infection.

Linos, Alexandros, Mahmoud M. Berekaa, Alexander Steinbüchel, Kwang Kyu Kim, Cathrin Spöer, and Reiner M. Kroppenstedt. "*Gordonia westfalica* sp. nov., a Novel Rubber-degrading Actinomycete." *International Journal of Systematic and Evolutionary Microbiology* 52 (2002): 1133-39. http://ijs.sgmjournals.org/cgi/reprint/52/4/1133.pdf.

Mullis, Kary. "The Polymerase Chain Reaction." Nobel Prize lecture, December 8, 1993. http://nobelprize.org/nobel_prizes/chemistry/laureates/1993/mullis-lecture.html.

Murray, John F. "A Century of Tuberculosis." *American Journal of Respiratory and Critical Care Medicine* 169 (2004): 1184-86. http://ajrccm.atsjournals.org/cgi/content/full/169/11/1181#FIG4.

Rose, Karsten, and Alexander Steinbüchel. "Biodegradation of Natural Rubber and Related Compounds: Recent Insights into a Hardly Understood Catabolic Capability of Microorganisms." *Applied and Environmental Microbiology* 71 (2005): 2803-12. http://aem.asm.org/cgi/content/full/71/6/2803#T1.

Stephenson, Shauna. "Thermus aquaticus." *Wyoming Tribune Eagle*, August 17, 2007. http://www.wyomingnews.com/articles/2007/08/17/outdoors/01out_08-15-07.txt.

## Chapter 5

### *Print*

Lovelock, James E. *Gaia: A New Look at Life on Earth*. Oxford University Press, 2000.

### *Internet*

Chung, King-Thom, and Christine L. Case. "Sergei Winogradsky: Founder of Soil Microbiology." *Society for Industrial Microbiology News* 51 (2001): 133-35. http://www.skylinecollege.edu/case/envmic/winogradsky.pdf.

Gemerden, Hans van. "Diel Cycle of Metabolism of Phototrophic Purple Sulfur Bacteria in Lake Cisó (Spain)." *Limnology and Oceanography* 30 (1985): 932-43. http://www.jstor.org/pss/2836576.

Guerrero, Ricardo, Carlos Pedrós-Alió, Isabel Esteve, Jordi Mas, David Chase, and Lynn Margulis. "Predatory Prokaryotes: Predation and Primary Consumption Evolved in Bacteria." *Proceedings of the National Academy of Sciences* 83 (1986): 2138-42. http://www.ncbi.nlm.nih.gov/pmc/articles/PMC323246/pdf/pnas00311-0181.pdf.

Pedrós-Alió, Carlos, Emilio Montesinos, and Ricardo Guerrero. "Factors Determining Annual Changes in Bacterial Photosynthetic Pigments in Holomictic Lake Cisó, Spain." *Applied and Environmental Microbiology* 46 (1983): 999-1006. http://aem.asm.org/cgi/reprint/46/5/999.

## Chapter 6

### *Print*

Amici, A., M. Bazzicalupo, E. Gallori, and F. Rollo. "Monitoring a Genetically Engineered Bacterium in a Freshwater Environment by Rapid Enzymatic Amplification of Synthetic DNA 'Number-plate.'" *Applied Microbiology and Biotechnology* 36 (1991): 222-27.

Emiliani, Cesare. *Planet Earth. Cosmology, Geology, and the Evolution of Life and Environment*. Cambridge University Press, 1992.

Gold, Thomas. *The Deep Hot Biosphere*. Springer-Verlag, 1999.

Meckel, Richard A. *Save the Babies: American Public Health Reform and the Prevention of Infant Mortality, 1850-1929*. Johns Hopkins University Press, 1990.

Rifkin, Jeremy. *The Biotech Century: Harnessing the Gene and Remaking the World*. Tarcher/Putnam, 1998.

Robbins-Roth, Cynthia. *From Alchemy to IPO*. Perseus Publishing, 2000.

### Internet

Anderson, A. J., and E. A. Dawes. "Occurrence, Metabolism, Metabolic Role, and Industrial Uses of Polyhydroxyalkanoates." *Microbiological Reviews* 54 (1990): 450-72. http://www.ncbi.nlm.nih.gov/pmc/articles/PMC372789.

Brand, David. "Gold Finds Our Deep Hot Biosphere Teeming with Life—And Controversy." *Cornell Chronicle*, January 28, 1999. http://www.news.cornell.edu/chronicle/99/1.28.99/Gold-book.html.

Budsberg, K. J., C. F. Wimpee, and J. F. Braddock. "Isolation and Identification of *Photobacterium phosphoreum* from an Unexpected Niche: Migrating Salmon." *Applied and Environmental Microbiology* 69 (2003): 6938-42. http://www.ncbi.nlm.nih.gov/pmc/articles/PMC262280/#r20.

Dagert, M., and S. D. Ehrlich. "Prolonged Incubation in Calcium Chloride Improves the Competence of *Escherichia coli* Cells." *Gene* 6 (1979): 23-38. http://www.sciencedirect.com/science?_ob=ArticleURL&_udi=B6T39-47W0K6H-53&_user=10&_rdoc=1&_fmt=&_orig=search&_sort=d&_docanchor=&view=c&_searchStrId=1080165555&_rerunOrigin=google&_acct=C000050221&_version=1&_urlVersion=0&_userid=10&md5=4e6ef6491becd14975cd51be5be6f7be.

Food and Agricultural Organization of the United Nations. "Hydrogen Production." Chap. 5 in *Renewable Biological Systems for Alternative Sustainable Energy Production*. Edited by Kasuhisha Miyamoto. FOA, 1997. http://www.fao.org/docrep/w7241e/w7241e0g.htm#5.2%20biophotolysis%20of%20water%20by%20microalgae%20and%20cyanobacteria.

Garcia, Belén, Elías R. Olivera, Baltasar Miñambres, Martiniano Fernández-Valverde, Librada M. Cañedo, María A. Prieto, José L. Garcia, María Martínez, and José M. Luengo. "Novel Biodegradable Aromatic Plastics from a Bacterial Source." *Journal of Biological Chemistry* 41 (1999): 29228-41. http://www.jbc.org/content/274/41/29228.full.pdf.

*Geographical*. "Biotech Could Make Chemical Production Carbon Neutral." March 2008. http://findarticles.com/p/articles/mi_hb3120/is_3_80/ai_n29416979.

Human Genome Project. http://www.ornl.gov/sci/techresources/Human_Genome/project/about.shtml.

Irrgang, Karl, and Ulrich Sonnenborn. *The Historical Development of Mutaflor Therapy*. Herdecke, Germany: Ardeypharm GMBH, 1988. http://www.ardeypharm.de/pdfs/en/mutaflor_historical_e.pdf.

Kelly, Michael. "Earthly Cave Bacteria Hint at Mars Life." *Discovery News*, May 8, 2009. http://dsc.discovery.com/news/2009/05/08/cave-bacteria-mars.html.

Kotlar, Hans Kristian. "Can Bacteria Rescue the oil Industry?" *The Scientist* 23 (2009): 30. http://www.the-scientist.com/article/display/55375/;jsessionid=6CE545DF031C7AC0C35910887AB34FC8.

Mandel, M., and A. Higa. "Calcium-dependent Bacteriophage DNA Infection." *Journal of Molecular Biology* 53 (1970): 159-62.

Shulman, Stanford T., Herbert C. Friedmann, and Ronald H. Sims. "Theodor Escherich: The First Pediatric Infectious Diseases Physician?" *Clinical Infectious Diseases* 45 (2007): 1025-29. http://www.journals.uchicago.edu/doi/pdf/10.1086/521946?cookieSet=1.

Singh, Mamtesh, Sanjay K. S. Patel, and Vipin C. Kalia. "*Bacillus subtilis* as a Potential Producer for Polyhydroxyalkanoates." *Microbial Cell Factories* 8 (2009): 38-49. http://www.microbialcellfactories.com/content/pdf/1475-2859-8-38.pdf.

Society for General Microbiology. "E. coli K-12: Joshua Lederberg." *Microbiology Today* 31, August 2004. http://www.sgm.ac.uk/pubs/micro_today/pdf/080402.pdf.

Williams, David R. "Evidence of Ancient Martian Life in Meteorite ALH84001." NASA, January 9, 2005. http://nssdc.gsfc.nasa.gov/planetary/marslife.html.

## Chapter 7

### Print

Deffeyes, Kenneth S. *Hubbert's Peak: The Impending World Oil Shortage*. Princeton, NJ: Princeton University Press, 2001.

Hackstein, Johannes H. P., and Claudius K, Stumm. "Methane Production in Terrestrial Arthropods." *Proceedings of the National Academy of Sciences* 71 (1994): 5441-45.

Lynn, Denis H. *The Ciliated Protozoa: Characterization, Classification, and Guide to the Literature*. Springer Science, 2008.

Reisner, Erwin, Juan C. Fontecilla-Camps, and Fraser A. Armstrong. "Catalytic Electrochemistry of a [NiFeSe]-hydrogenase on $TiO_2$ and Demonstration of Its Suitability for Visible Light-driven $H_2$ Production." *Chemical Communications* 7 (2009): 550-52.

Shah, Sonia. *Crude: The Story of Oil*. New York: Seven Stories Press, 2004.

### Internet

Charlson, Robert J., James E. Lovelock, Meinrat O. Andreae, and Stephen G. Warren. "Oceanic Phytoplankton, Atmospheric Sulphur, Cloud Albedo and Climate." *Nature* 326 (1987): 655-61.

Green Car Congress. "Researchers Develop Bacterial Enzyme-Based Catalyst for Water-Gas Shift Reaction at Ambient Conditions; New Thinking About Catalyst Design." News release, September 22, 2009. http://www.greencarcongress.com/2009/09/rbio-wgs-20090922.html.

Harten, Alan. "Bacteria that Makes Hydrogen Fuel." *Fair Home*, January 15, 2009. http://www.fairhome.co.uk/2009/01/15/bacteria-that-makes-hydrogen-fuel.

Henstra, Anne M., and Alfons J. M. Stams. "Novel Physiological Features of *Carboxydothermus hydrogenoformans* and *Thermoterrabacterium ferrireducans*." *Applied and Environmental Microbiology* 70 (2004): 7236-40. http://www.ncbi.nlm.nih.gov/pmc/articles/PMC535181.

Highlights in Chemical Technology. "Sun Shines on a Solution for Hydrogen Production." December 8, 2008. http://www.rsc.org/Publishing/ChemTech/Volume/2009/02/sun_shines_hydrogen.asp.

Kiene, Ronald P. "Dimethyl Sulfide Production from Dimethylsulfoniopropionate in Coastal Seawater Samples and Bacterial Cultures." *Applied And Environmental Microbiology* 56 (1990): 3292-97.

Ocampo, R., H. J. Callot, and P. Albrecht. "Evidence of Porphyrins of Bacterial and Algal Origin in Oil Shale." Chap. 3 in *Metal Complexes in Fossil Fuels*. American Chemical Society, 1997. http://pubs.acs.org/doi/abs/10.1021/bk-1987-0344.ch003.

PhysOrg.com. "Food from Fuel Waste: Bacteria Provide Power." http://www.physorg.com/news135482832.html.

Rasmussen, Birger, Tim S. Blake, Ian R. Fletcher, and Matt R. Kilburn. "Evidence for Microbial Life in Synsedimentary Cavities from 2.75 Ga Terrestrial Environments." *Geology* 37 (2009): 423-26.

Rawal, B. D., and A. M. Pretorius. "*Nanobacterium sanguineum*—Is it a New Life-form in Search of Human Ailment or Commensal: Overview of Its Transmissibility and Chemical Means of Intervention." *Medical Hypotheses* 65 (2005): 1062-66. http://www.ncbi.nlm.nih.gov/pubmed/16122881.

Savage, Neil. "Making Gasoline from Bacteria." *MIT Technology Review*, August 1, 2007. http://www.technologyreview.com/Biztech/19128.

Schaechter, Moselio, John L. Ingraham, and Frederick C. Neidhardt. *Microbe*. American Society for Microbiology Press, 2006.

*ScienceDaily*. "Fuel from Bacteria is One Step Closer." August 8, 2008. http://www.sciencedaily.com/releases/2008/08/080806113141.htm.

SpaceRef Interactive. "NASA's Johnson Space Center to Study Nanobacteria." Press release, September 13, 2004. http://www.spaceref.com/news/viewpr.html?pid=15024.

U.S. Environmental Protection Agency. http://www.epa.gov.

Zimmer-Faust, Richard K., Mark P. de Souza, and Duane C. Yoch. "Bacterial Chemotaxis and Its Potential Role in Marine Dimethylsulfide Production and Biogeochemical Sulfur Cycling." *Limnology and Oceanography* 41 (1996): 1330-34.

ZoBell, Claude E. "Contributions of Bacteria to the Origin of Oil." Presented at World Petroleum Congress, The Hague, The Netherlands, May 28-June 6, 1951. http://www.onepetro.org/mslib/servlet/onepetropreview?id=WPC-4029&soc=WPC&speAppNameCookie=ONEPETRO.

# Index

## b

## C

## d



## e

## n

## q–r

## S

## x–y–z

FT Press
FINANCIAL TIMES

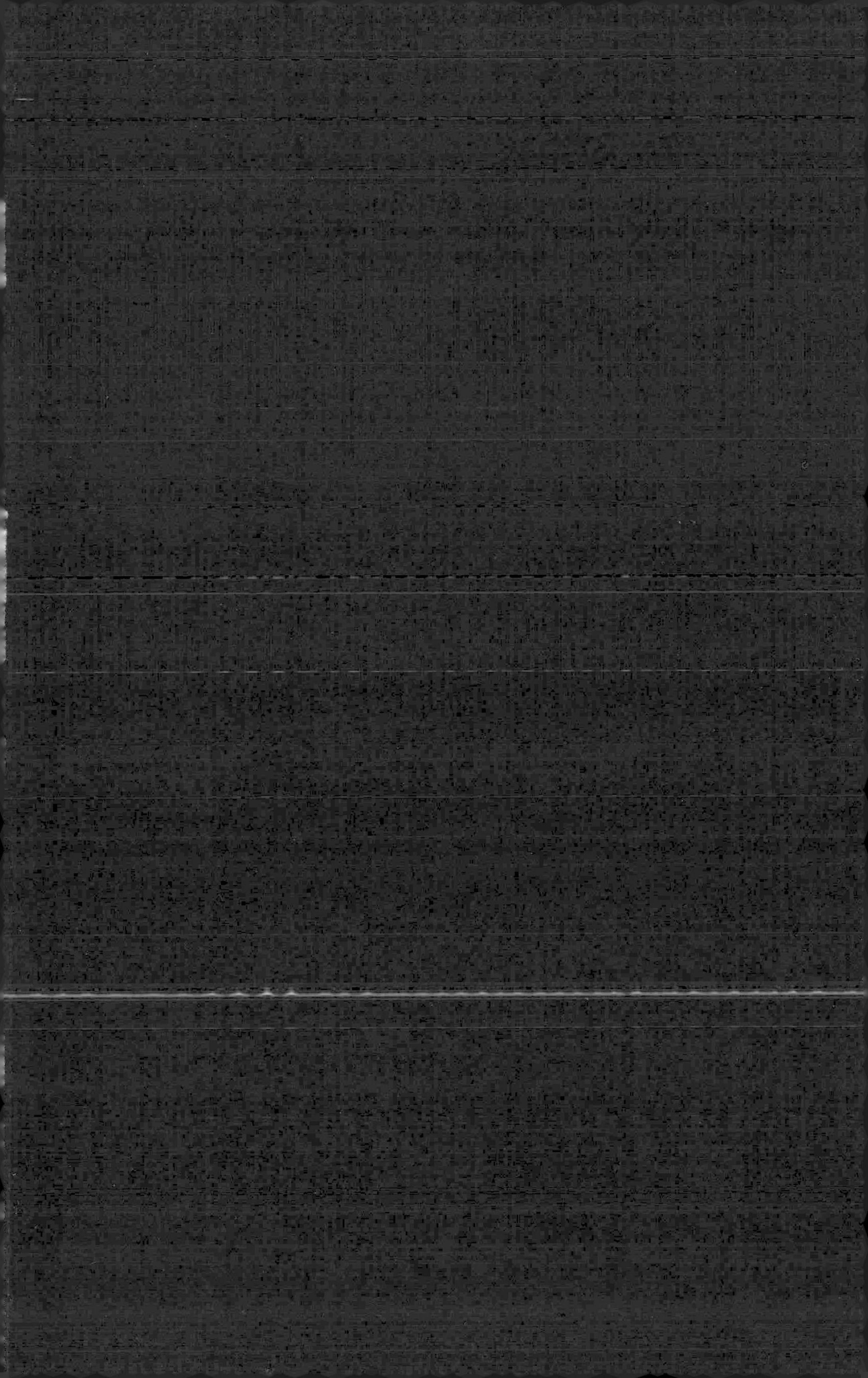